Cultures of Care in Irish Medical History, 1750–1970

Cultures of Care in Irish Medical History, 1750–1970

Edited By

Catherine Cox
Director, Centre for the History of Medicine, University College, Dublin

Maria Luddy
Professor of Modern Irish History, University of Warwick

First published 2010 by
PALGRAVE MACMILLAN

Palgrave Macmillan in the UK is an imprint of Macmillan Publishers Limited,
registered in England, company number 785998, of Houndmills, Basingstoke,
Hampshire RG21 6XS.

Palgrave Macmillan in the US is a division of St Martin's Press LLC, 175 Fifth
Avenue, New York, NY 10010.

Palgrave Macmillan is the global academic imprint of the above companies
and has companies and representatives throughout the world.

Palgrave® and Macmillan® are registered trademarks in the United States,
the United Kingdom, Europe and other countries

ISBN 978-0-230-53586-2   hardback

This book is printed on paper suitable for recycling and made from fully
managed and sustained forest sources. Logging, pulping and manufacturing
processes are expected to conform to the environmental regulations of the
country of origin.

A catalogue record for this book is available from the British Library.

Library of Congress Cataloging-in-Publication Data
    Cultures of care in Irish medical history, 1750–1970 / edited by
    Catherine Cox, Maria Luddy.
        p. cm.
    Includes bibliographical references and index.
    ISBN 978-0-230-53586-2 (hardback)
        1. Medical–Ireland–History. 2. Medical care–Ireland–History.
    I. Cox, Catherine, 1970– II. Luddy, Maria.
    [DNLM: 1. Delivery of Health Care–history–Ireland. 2. History,
    18th Century–Ireland. 3. History, 19th Century–Ireland.
    4. History, 20th Century–Ireland. 5. Social Conditions–history–Ireland.
    WZ 70 GI6 C898 2010]
    R498.6.C85 2010
    362.109417–dc22   2010023757

10   9   8   7   6   5   4   3   2   1
19   18 17 16 15 14 13 12 11 10

Contents

List of Tables   vii

Acknowledgements   viii

Notes on Contributors   ix

Introduction   1
Catherine Cox and Maria Luddy

1   'Bleeding, vomiting and purging': The Medical   13
    Response to Ill-health in Eighteenth-century Ireland
    James Kelly

2   General Practice and Coroners' Practice: Medico-legal   37
    Work and the Irish Medical Profession, c. 1830–c. 1890
    Michael J. Clark

3   Access and Engagement: The Medical Dispensary Service   57
    in Post-Famine Ireland
    Catherine Cox

4   Suicide and Insanity in Post-Famine Ireland   79
    Georgina Laragy

5   Psychiatry and the Fate of Women Who Killed Infants   92
    and Young Children, 1850–1900
    Pauline M. Prior

6   Science, Politics and the Irish Literary Revival: 113
    Reassessing 'Dr Sigerson' as Polymath and Public
    Intellectual
    James McGeachie

7   'This Revived Old Plague': Coping with Flu   141
    Caitríona Foley

8   'Half mad at the time': Unmarried Mothers and   168
    Infanticide in Ireland, 1922–1950
    Clíona Rattigan

9   Venereal Disease in Interwar Northern Ireland   191
    Leanne McCormick

10 Moral Prescription: The Irish Medical Profession,   207
    the Roman Catholic Church and the Prohibition of
    Birth Control in Twentieth-century Ireland
    Lindsey Earner-Byrne

11 Death and Disease in Independent Ireland,   229
    c. 1920–1970: A Research Agenda
    Mary E. Daly

Index   251

List of Tables

5.1   Women who killed an infant (age under one year)   96
        and found to be insane at the time of the crime
5.2   Women who killed a child (age over one year) and   97
        found to be insane at the time of the crime
11.1   Mean number of children born per 100 women   229
        with marriages of 20–24 years' duration
11.2   Life expectancy at birth in Ireland   231
11.3   Life expectancy at birth in various European   232
        countries
11.4   Major infectious diseases, 1940–44   234

Acknowledgements

The editors would like to thank the Centre for the History of Medicine at the University of Warwick for supporting the original workshop at which these papers were presented. We are also grateful to the Wellcome Trust which funded that workshop. We made use of a wide network of colleagues to review the articles in this book and we would like to thank them for their comments and support for this publication.

Catherine Cox and Maria Luddy

Notes on Contributors

Michael J. Clark is a Visiting Lecturer attached to the Centre for the Humanities and Health in the Department of English Language and Literature, King's College London, where he teaches courses on medicine and film. Between 2004 and 2008 he was an Honorary Research Associate of the Wellcome Trust Centre for the History of Medicine at University College London, where he worked on medico-legal practice in nineteenth and early twentieth-century Ireland, in particular the role of the Irish medical profession in Irish coroners' practice. Dr Clark has published a number of articles and essays on various aspects of the history of psychiatry and legal medicine, and the representation of medical practice and bio-medical science in fiction and films. He is also co-editor with Catherine Crawford of *Legal Medicine in History* (1994).

Catherine Cox is a Lecturer at the School of History and Archives, University College Dublin (UCD), and is also the UCD Director of the Wellcome Trust Centre for the History of Medicine in Ireland. She has published articles and book chapters on medical practitioners and practice in the nineteenth century and on institutionalisation in Ireland. She has also worked on the history of insanity and asylums in nineteenth-century Ireland (*Managing Insanity in Nineteenth-Century Ireland* (forthcoming, 2010)). With Prof. Hilary Marland, University of Warwick, she is currently working on a Wellcome Trust funded project that is investigating the relationship between Irish migrants, ethnicity and mental illness from c.1850 to 1921.

Mary E. Daly is Principal of the College of Arts and Celtic Studies, University College Dublin, and a Professor of Irish History. From 2000–2004 she served as Secretary of the Royal Irish Academy; she is deputy-chair of the Higher Education Authority. Educated at UCD and Nuffield College Oxford, she has held visiting positions at Harvard and Boston College and is a former member of the National Archives Advisory Council and the Irish Manuscripts Commission. She is a member of the Centre for the History of Medicine in Ireland, which is funded by the Wellcome Trust. Her research interests are concentrated on Ireland in the nineteenth and twentieth centuries, primarily (though not exclusively) in socio-economic history. Recent publications include *The Slow*

Failure: Population Decline and Independent Ireland 1920–1973 (2006), which was nominated as a Choice Outstanding Academic Title for 2006; *1916 in 1966: Commemorating the Easter Rising*, co-edited with Margaret O'Callaghan (2007), and '"The primary and natural educator"? The role of parents in the education of their children in independent Ireland', *Eire-Ireland*, 44 (spring/summer 2009).

Lindsey Earner-Byrne has lectured in Modern Irish History in the School of History and Archives, University College Dublin since 2001. She has published in the area of gender and family history. Her book *Mother and Child: Maternity and Child Welfare in Dublin, 1922–60* was published in 2007. She has also published several journal articles and book chapters. She is the co-ordinator of the Evolution of the Irish States' Project at UCD. The current phase of the project: 'The Evolution of the Irish State Project 2008–2011: Charity Cases in Modern Ireland, 1921–39' is focused on the poor of modern Ireland. This project seeks to provide a comprehensive analysis of poverty and charity during the first two decades of Irish independence.

Caitríona Foley recently completed a PhD in History at University College Dublin, and was a Government of Ireland Postgraduate Scholar. Her thesis examined the ways in which the Great Flu pandemic of 1918–19 affected Ireland, with particular emphasis on its social and cultural dimensions. She holds a BA from Trinity College Dublin, where she studied German and History, and has also worked as a researcher at the Centre of the History of Medicine in Ireland.

James Kelly, M.R.I.A., is Head of the History Department at St Patrick's College, Dublin City University, and Honorary Research Professor in History at Queen's University Belfast. He has published widely on the social, political and religious history of late early modern Ireland. His recent publications include *The Liberty and Ormond Boys: Faction Fighting in Eighteenth-Century Dublin* (2005), *Poynings' Law and the Making of Law in Ireland, 1660–1800* (2007), *The Proceedings of the Irish House of Lords, 1771–1800* (3 vols, 2008), and *Sir Richard Musgrave, 1746–1818: Ultra-Protestant Ideologue* (2009).

Georgina Laragy was an Irish Research Council for the Humanities and Social Sciences (IRCHSS) Government Scholar at the National University of Ireland (NUI), Maynooth, from 2001 to 2004, during which time she completed her doctoral thesis, 'Suicide in Ireland,

1831–1921: a Social and Cultural Study'. This study is currently being transformed into a book. Since 2004 she has worked as a lecturer at NUI Maynooth and as a Postdoctoral Research Assistant on an Economic and Social Research Council (ESRC) funded project 'Welfare Regimes under the Irish Poor Law, 1850–1920' based at Oxford Brookes University. She is currently working on an IRCHSS funded project based at the University of Limerick entitled 'From the Cradle to the Grave: Lifecycles in Modern Ireland'. Her interests include the history of suicide, poverty and medicine and she is particularly interested in exploring the intersections between criminal and medical history as well as the relationship between institutions and their communities. Her publications include 'Suicide among the Irish and English emigrants in 1870s New York', *Proceedings of the Associação Portuguesa de Estudos Anglo-Americanos, XXVII Encontro, 'Crossroads of History and Culture' I*, 2007 and 'Murder in Cavan, 1809–1891', *Breifne: Journal of Cumann Seanchais Bhréifne*, 44 (2008).

Maria Luddy is Professor of Modern Irish History at the University of Warwick. She has published extensively on Irish social history, particularly the history of women. Recent publications include *Prostitution and Irish Society, 1800–1940* (2007). She is currently co-directing an Arts and Humanities Research Council (AHRC) project on the history of marriage in Ireland.

Leanne McCormick is Lecturer in Modern Irish Social History at the University of Ulster. Her research and publications have considered the attempts to control female sexuality in Northern Ireland and her monograph, *Regulating Sexuality: Women in Twentieth-Century Northern Ireland* was published in 2009. Her present research focuses on female health in twentieth-century Northern Ireland, with particular regard to the issues of family planning and abortion. She is also working on a project considering the various institutions to which 'troublesome' girls were sent in Belfast in the first half of the twentieth century. She is a member of the Centre for the History of Medicine in Ireland, which is funded by the Wellcome Trust.

James McGeachie teaches at Our Lady and St Patrick's College, Belfast and is an Associate Member of the Academy of Irish Cultural Heritages at the University of Ulster. He is a graduate of Reading University, with an MA from the University of Manitoba (Canada) and a PhD from Cambridge. He taught previously at the Jews' Free School and Hampstead School in London and at Cambridge University, University College

London, the University of Notre Dame (Indiana) and the University of Ulster, where he was also a Wellcome Research Fellow. His teaching and research interests are chiefly in Irish and British cultural and intellectual history and literature since the eighteenth century and the history of medicine. He has published studies of the uses of complexity in late-Victorian writing, of Sir William Wilde and the nineteenth-century Dublin school of medicine, 'Arthur Balfour's Cambridge and the 1870s reaction against Mill', and entries on nineteenth-century Irish and British clinicians in W.F. Bynum and H. Bynum's 2007 *Dictionary of Medical Biography*.

Pauline M. Prior teaches Social Policy at Queen's University Belfast. With a background in social work, her research focuses on different aspects of mental health policy, including gender, law and history. Her articles have appeared in a variety of journals including *Eire-Ireland* and *New Hibernia Review*. She has published five books, the latest of which is *Madness and Murder: Gender, Crime and Mental Disorder in Nineteenth-Century Ireland* (2008).

Clíona Rattigan is a Teaching Fellow at the University of Warwick where she lectures in modern Irish history. Her PhD, funded by the IRCHSS, was completed in the Department of History, Trinity College Dublin, in October 2007. Her research interests and publications centre on single motherhood in twentieth-century Ireland, infanticide and abortion cases. She is currently working on a study of single mothers and infanticide in Ireland during the first half of the twentieth century. Her publications include: '"Done to death by father or relatives": Irish families and infanticide cases, 1922–1950', *History of the Family: An International Quarterly*, 13:4 (2008) and '"Crimes of passion of the worst character": abortion cases and gender in Ireland, 1922–1950', in Maryann Valiulis (ed.), *Gender and Power in Irish History* (2008).

Introduction

Catherine Cox and Maria Luddy

'How societies organise health care, how individuals or states relate to sickness, how we understand our identity and agency as sufferers or healers – [these questions] are simply too important for the practices of medical history not to be persistently subjected to vigorous reflection and re-examination.'[1]

In the course of the last five years or more, historians of medicine have engaged in a timely assessment of past and future intellectual developments within the history of medicine. In part, this has been prompted by the twentieth anniversary of the Society for the Social History of Medicine, one of the main societies devoted to the field of medical history. The Society's journal, *Social History of Medicine*, dedicated its 2007 edition to a serious reflection on the current state of the history of medicine and its future.[2] This debate on the future of the history of medicine has also been inspired by the new intellectual directions evident within the field of history as a whole over the last two decades and by evolving political, economic and cultural contexts. A consensus has currently emerged among a group of scholars that the history of medicine is lacking 'fresh theoretical engagement and analysis'.[3] The 'taking-stock' of the history of medicine is not confined to the Anglo-American sphere. For example, Frank Huisman and John Harley Warner's edited collection *Locating Medical History: The Stories and Their Meanings* attempts to bring a European dimension to the discussion.[4] That volume includes essays on some of the 'historiographical and ideological issues' that have preoccupied medical historians in France and Germany, including the 'fall' of medical history in Germany and the importance of positivist medical history in France.[5] A stated aim of the collection was to 'ruthless[ly] look at the practise of medical history in

1

the present, recognizing diversity in historians' backgrounds, approaches, aspirations, and audiences' and some of the essays are successful in this endeavour.[6]

Concomitant with the emergence of this debate, the most significant funder of medical history, the Wellcome Trust, has gone through a process of review and re-assessment. This culminated in the announcement in May 2009 that the Wellcome Trust was realigning its focus to include not just 'medical history' but also 'medical history and the humanities'. While definitions of the medical humanities are contested, scholars portray it as a challenge to medicine 'to become interdisciplinary, and be disciplined by arts and humanities as well as science'.[7] The Wellcome Trust has stressed that the analytical tools of the field of history should underpin the medical humanities. The intention behind the Wellcome Trust's shift in focus is to encourage 'applicants to address the important questions that will develop further our understanding of the progress, socio-economic and cultural impacts of medicine and medical sciences on human and animal health'.[8] The explicit focus on the medical humanities within the funding activities of the Wellcome Trust has led to the establishment of a Centre for the Humanities and Health at King's College, London, and appointments in medical humanities. The Trust has also drawn applicants' attention to the interactions between history and current public policy, and the importance of communicating the history of medicine and medical humanities to the general public and policy makers. The entreaty to broaden the remit of medical history is not confined to the Wellcome Trust's interest in the medical humanities. Scholars have decried 'single site studies' and the failure to 'analyse events from a comparative perspective'.[9] As in other areas of history, and indeed society, we are encouraged to think globally.

These discussions have taken place at a time when Irish medical history has entered a period of greater vibrancy. In 1999 Greta Jones and Elizabeth Malcolm brought together the work of scholars writing on aspects of the social history of medicine in Ireland in an important edited collection, *Medicine, Disease and the State in Ireland 1650–1940*.[10] The editors' introduction to the collection is a comprehensive overview of the field at the end of the millennium, and is an important contribution to the historiography of medical history in Ireland. The articles contained in the volume interrogate, among other issues, the evolution of institutional medicine in Ireland in terms of educational provision and the expansion of the hospital, the phenomenon of 'miraculous cures', the role of nurses in medical relief, and the problems that faced Irish medicine by the continued presence of diseases such as typhus, smallpox and tuberculosis. Some of these themes are echoed in this current collection.

Since 1999 a number of important scholarly contributions to the field have been published,[11] and articles exploring aspects of the medical history of Ireland have started to appear with greater frequency in the main journals.[12] The majority of these publications have consciously or unconsciously responded to Jones' and Malcolm's call for medical historians of Ireland to move away from institutional histories and from positivistic, factual, descriptive approaches and to place their work in a broader context of medical or Irish history.[13] Also since 1999, there continues to be a healthy number of graduate students studying Irish medical history, in Irish and British universities, and this was particularly evident during the 2009 international postgraduate conference of the Society for the Social History of Medicine. Recently completed doctoral theses have broken new ground[14] and an MA in the topic has been established.[15] Conferences and workshops devoted to the subject and associated fields frequently feature in the academic calendar.

The current vibrancy in the field of Irish medical history is in part due to the emergence in the last ten years or so of a number of projects and initiatives in receipt of substantial external funding. In 2006 the Centre for the History of Medicine in Ireland was founded on receipt of a Wellcome Trust Strategic Enhancement Award. The Centre is a collaborative project between the School of History and Archives at University College Dublin and the School of History and International Affairs at the University of Ulster, with the expressed aim of developing the discipline of medical history in Ireland in terms of both teaching and research.[16] At Trinity College, Dublin, the Irish Research Council for the Humanities and Social Sciences has funded a project on 'Ireland, Empire and Education' that is investigating the role Irish graduates played in shaping imperial and colonial processes, particularly in the fields of engineering, medicine, law and theology between 1840 and 1940. An ongoing Economic and Social Research Council funded project at Oxford Brookes University, led by Dr Virginia Crossman, examines the development of the poor law in Ireland from the end of the Great Famine to the establishment of the Irish Free State, in the context of welfare and health issues. In addition, existing societies, such as the Eighteenth-Century Ireland Society, and the Society for the Study of Nineteenth Century Ireland have organised conferences on medical history. The proceedings of the Eighteenth-Century Ireland Society annual conference held in 2008 have recently been published.[17]

In the context of the re-assessment of medical history internationally, and the repeated calls for the broadening of the 'medical history remit', a volume devoted to Irish medical history may seem out of step with the intellectual climate. However, in many respects, the increased interest in the subject in Ireland at a time of reflection could be extremely fortuitous and encouraging. Local histories of medicine can ask global questions. In many respects medicine is, and has been, a global endeavour – for example, there are certain commonalities in terms of treatment, education, experiences and knowledge – though scholars must be wary of exaggerating the similarities. As Ilana Löwy has observed

> Diseases are trans-national phenomena; medicine and public health are often global endeavours; medical practice is influenced by international rules and regulations, and by global economic and political trends; medical researchers and medical practitioners travel, as do their ideas; developments in one country may affect those in other countries.[18]

Simultaneously, the delivery of healthcare and medicine is shaped by local circumstances. Contributors to this volume have explored how religious, legal and 'traditional' practices, as well as state policy, shaped the development of birth control policies, attitudes towards suicide and infanticide, state and societal responses to disease, and the delivery of healthcare. When placed within a comparative framework, such studies can highlight the important continuities and departures from the broad history of medicine narrative and analysis.[19] This is not simply 'filling-in' the Irish case, but testing many of the concepts and the hermeneutic devices that medical historians deploy in their scholarship and teaching. It is also true that though historians sometimes exaggerate the 'exceptional' status claimed for the history of Ireland, there are instances where a comparative exploration does illuminate national distinctions. For instance, a recent study of the history of vaccination reveals that Ireland's vaccination policy, though originating in an 1840 Act that applied to England and Wales, was substantially different to that which emerged in the rest of the United Kingdom. The Vaccination Acts of 1858 and 1863 ensured that by the late 1860s more than 70 per cent of infants had been vaccinated[20] due to the use of the extensive dispensary system operating under the 1851 Medical Charities' Act, to deliver the service to the Irish poor.

In 1999, Jones and Malcolm identified important topics of research that warrant serious attention from historians of Irish medical history, such as medical associations, traditional and popular medicine, nursing, medical migration and pharmacy.[21] Since then, research has been completed on aspects of these topics and a number of studies have been published or are due to appear shortly.[22] A proportion of this work has emerged from recent postgraduates and remains as yet unpublished, although some of it is featured in this collection.[23] Recent innovations have also ensured that there is a greater awareness of the medical history sources held in Ireland and a number of preservation and cataloguing projects have been initiated.[24] The transfer of the archives of the Irish Nursing Board (An Bord Altranais) to University College Dublin Archives in July 2007 should prompt scholarly research into the history of Irish nursing and midwifery.[25] Generous funding from the Wellcome Trust has assisted research into medical migration.[26] The history of diseases continues to attract scholarly attention, with a growing literature in particular on venereal diseases in Ireland.[27] In 1999, Jones and Malcolm commented on the vibrancy of work on the 'lunatic asylum' in nineteenth-century Ireland and this interest has continued with the publication of new work by Catherine Cox, Brendan Kelly, Elizabeth Malcolm, Pauline Prior and Oonagh Walsh.[28] The history of psychiatry is also now attracting scholars exploring twentieth-century developments.[29]

Despite this vibrancy, certain aspects of Irish medical history remain relatively untouched. The work of Waltraud Ernst and others has highlighted the centrality of colonial medicine, yet few scholars of the history of medicine in Ireland have demonstrated a willingness to engage with the subject.[30] This is undoubtedly due to the disputed and complex nature of Ireland's status as a colony.[31] There is also a paucity of work on occupational health. A popular field of study in countries with a strong industrial history since the 1980s, occupational health has been more broadly conceived in recent years to include the health of agricultural workers.[32] In a predominantly rural country, such as Ireland, which also possessed industrialised centres, the opportunities for new innovative research in this field are strong. In a related area, there are also rich opportunities for exploring the history of animal health in Ireland and the cross-over with human diseases.[33] There is still little work on the medical care of children in Irish society. We know little about Irish attitudes to sick children, their care within the home or within institutions, and the influence of gender on Irish medical practice and the perception of patients. We also know little about how patients experienced their medical care and treatment, or how they understood the role of general practitioners or consultants. In this period of re-examination in the history of medicine, it is important that scholars of Irish medical history engage with the core debates and do not shy away from the admittedly vast international scholarship. Historians can now tap into fields of sophisticated medical history research that will not only illuminate how historians might tackle Irish medical history but will also allow for a greater degree of comparative history.

This collection of essays brings together papers first presented at a workshop in Warwick University in 2005.[34] The workshop was intended to gather together historians who research in the area of the history of medicine in Ireland to discuss their work and to explore comparative links within that history. The intention was to broaden the scope of Irish medical history and with this publication to bring this new research to a wider audience. The Ireland under discussion is, of course, made up of the State of Northern Ireland, created in 1920, and initially the Irish Free State (1922), later to become the Irish Republic from 1949. It is worth noting that legislation, which influenced or affected medical practice or healthcare, developed quite differently in both States, the most obvious example being the introduction of the National Health Service to Northern Ireland in 1949.

In keeping with the theme of this collection, the 'culture of medical care', the book begins in the eighteenth century where James Kelly combines an analysis of the nature of illness and its remedies with an exploration of the increased use and spread of proprietary medicines in the country. It was the 'moneyed class' who had the greatest access to proprietary medicines and the attentions of medical men. And the century saw a significant growth in the number of trained practitioners in Dublin in particular. James McGeachie's article on George Sigerson, an eminent figure in bio-medical science in Ireland and a public intellectual, shows how significant some medical men were to the Irish medical and cultural worlds. Sigerson introduced continental medical innovations to Ireland, held important medical chairs in the Catholic University School of Medicine, and contributed to public debates on issues of Irish national identity and culture. McGeachie argues that Sigerson was in a tradition of Irish medical men who debated in a specifically Irish context the relationships between science, modernity and national identity.

Michael Clark further explores the development of the medical profession through the evolution of the relationship between general medical practice and coroners' practice. The provision of medical evidence at coroners' inquests became an important issue for the Irish medical profession in the nineteenth century and was suffused with political and professional difficulties. Clark brings to light the similarities and differences that existed between the systems that operated in Ireland and England and elucidates how the medical profession utilised the debates surrounding coroners' practice to call for reform and change that eventually enhanced their own professional status.

Catherine Cox explores the development of medical provision for the poor in post-famine Ireland, looking especially at the dispensary system for the poor. Though historians have identified the dispensary service as an important element in the healthcare of the poor,[35] here Cox analyses the issues that affected patients' access to, and shaped the acceptance of, this service. For medical professionals, appointments as dispensary medical officers were keenly sought. These provided financial security, though problems did exist with practitioners holding a number of appointments at the same time. The distances doctors had to travel and the often difficult terrain they had to traverse made prompt attention to some patients problematic. Medical officers' relationships with, and attitudes to, local healers were subtle and nuanced, and not as dismissive as has been believed. Doctors, especially in rural Ireland, had to negotiate their status, not alone with the local population but also with the local clergy. Even up to the first decade of the twentieth century the free medical aid provided by the dispensary system was not always utilised by those who were ill, and Cox notes that the dispensary system remained a contested service.

Moving into the realm of mental health, Georgina Laragy, Pauline Prior and Cliona Rattigan explore the problematic subjects of suicide, and homicidal and infanticidal women. Exploring the legal limits and definitions of these subjects, these three authors illuminate the changing social and cultural contexts that saw suicide and infanticide medicalised, and they analyse how these subjects were understood more generally within Irish society. Laragy reveals that suicide was considered a crime in Ireland until 1993. In the nineteenth century the majority of suicides were judged to be temporarily insane and the Catholic Church tended to follow legal tradition in an effort to excuse the sin of self-murder. Such a belief in the link between insanity and suicide provided protection to the family from legal and ecclesiastical punishment. The evidence available from suicide cases shows that families were often concerned about the health of individuals who threatened and carried through on suicide attempts and tended to watch them closely. Likewise judges tended to be mindful of the welfare needs of attempted suicides. Prior, in an exploration of women who killed children, shows clearly that the link with insanity was a means used by general and medical society to understand the actions of such women. While many did not repent their deeds, numerous women were excused of responsibility on the grounds of insanity. Rattigan looks in detail at the role played by medical professionals in infanticide cases that were tried in the Central Criminal Court between 1922 and 1950. Recording the encounters between infanticidal women and medical professionals, Rattigan reveals the power that had become inherent in the medical profession by the 1920s. By examining the court records of these cases, Rattigan also notes the physical and emotional traumas endured by these women. Echoing Prior's findings for the nineteenth century, Rattigan shows the persistence of the idea that women suffering the trauma of childbirth could become temporarily insane. Attempted suicide by new mothers reinforced that link between insanity and childbirth. The passing of the 1949 Infanticide Act legalised the view that a woman charged with infanticide was no longer considered responsible for her actions.

Lindsey Earner-Byrne reveals how the issue of birth control, which might have changed the fate of many women who committed infanticide had knowledge been available to them, was consciously avoided in maternity care in Ireland. She argues that an alliance existed between Catholic doctors and the Catholic Church, after the foundation of the state, to ensure the Catholic nature of medicine in Ireland. Politics played a key role in the provision of medical care to women, and both the Catholic Church and the medical profession deemed it inappropriate to provide birth control information to potential mothers. Fearing the imposition of socialised medicine and state control, the alliance forged between the Irish medical profession and the Catholic Church in twentieth-century Ireland was essentially one that preserved their best interests.

Mary Daly, in an examination of mortality in Ireland from the 1920s to 1970, shows the significance of Ireland's demographic and socio-economic contexts to the health of the population. Placing her research in a comparative framework she looks specifically at the physical conditions, and the political and economic developments that shaped Irish mortality. The variety of diseases that assailed the Irish people were similar to those that affected people in Western Europe. Perhaps one of the most deadly of these diseases was the highly contagious influenza, which, over the centuries, has caused the deaths of millions of people. This disease struck Ireland, and the rest of the world, in force between 1918 and 1920. Caitriona Foley looks at this major epidemic, and the 1890 epidemic that preceded it, focusing in particular on how lay and medical communities dealt with and understood the disease. The home remained the main site of treatment for those struck down with influenza, and the commercial world provided numerous potions to ease the symptoms or promote a cure, revealing the increasing commodification of health in this period, maintaining, as Kelly's article shows, a trend evident in Ireland in the eighteenth century. Some diseases were not given much publicity. Amongst these were the venereal diseases, explored for the interwar period in Northern Ireland by Leanne McCormick. Even the medical profession exhibited a degree of reticence in discussing these diseases in Northern Ireland, a reticence echoed in the south of Ireland,[36] and venereal disease remained a low priority in the field of public health initiatives.

All of these chapters help to reveal the diversity of medical provision in Ireland from the eighteenth century. They also show how politics, economics and cultural beliefs shaped that provision. Central to this collection are the ways in which the medical community endeavoured to evolve into a professional community, through its commercial links, legal connections or alliances with powerful forces such as the Catholic Church. These chapters also identify how medicine attempted to meet the needs of the people, as well as the role that families played in caring for their ill relatives. Formal medicine was only gradually accepted by the Irish and, sometimes, like formal religion, coexisted with alternative beliefs and understandings about health and illness.

Notes

1 Frank Huisman and John Harley Warner, 'Medical histories', in idem. (eds), *Locating Medical History: The Stories and their Meanings* (Baltimore and London, 2004), p. 3.
2 The articles of particular note are Roger Cooter, '"After Death/After-Life": The social history of medicine in post-postmodernity', *Social History of Medicine*, 20:3 (2007), pp. 441–64; Flurin Condrau, 'The patient's view meets the clinical gaze', *Social History of Medicine*, 20:3 (2007), pp. 525–40; Ilana Löwy, 'The social history of medicine: beyond the local', *Social History of Medicine*, 20:3 (2007), pp. 465–81; Waltraud Ernst, 'Beyond east and west. From the history of colonial medicine to a social history of medicine(s) in South Asia', *Social History of Medicine*, 20:3 (2007), pp. 505–24.
3 Huisman and Warner, 'Medical histories', p. 2; Roger Cooter, '"Framing" the end of the social history of medicine', in Huisman and Warner (eds), *Locating Medical History*, pp. 309–37.
4 Huisman and Warner, *Locating Medical History*.
5 Bill Luckin, 'Review of Frank Huisman and John Harley Warner (eds), *Locating Medical History: The Stories and Their Meanings* (Baltimore and London, 2004)', *Medical History*, 50:1 (2006), p. 139.

6 Huisman and Warner, *Locating Medical History*, p. 3. See also Luckin, 'Review of Frank Huisman and John Harley Warner', pp. 139–41 and Ed Morman, 'Review of Frank Huisman and John Harley Warner (eds), *Locating Medical History: The Stories and Their Meanings* (Baltimore and London, 2004)', *Journal of The History of Medicine and Allied Sciences*, 62:1 (2007), pp. 115–17.

7 Gillie Bolton, 'Boundaries of humanities: Writing medical humanities', *Arts and Humanities in Higher Education*, 7:2 (2008), pp. 131–48.

8 Presentation by Liz Shaw and Nils Fietje, 'Medical humanities at the Wellcome Trust', Trinity College Dublin, 27 May 2009. For more details on the Wellcome Trust guidelines see www.wellcome.ac.uk/Funding/Medical-humanities/index.htm.

9 Löwy, 'Social history of medicine: Beyond the local', p. 466.

10 Greta Jones and Elizabeth Malcolm (eds), *Medicine, Disease and the State in Ireland, 1650–1940* (Cork, 1999).

11 For example see Greta Jones, *'Captain of all these Men of Death': The History of Tuberculosis in Nineteenth and Twentieth-Century Ireland* (Amsterdam, 2001); L.A. Clarkson and E. Margaret Crawford, *A History of Food and Nutrition in Ireland 1500–1920* (Oxford, 2001); Tony Farmar, *Patients, Potions and Physicians: Social History of Medicine in Ireland* (Dublin, 2004); Gary A. Boyd, *Dublin, 1745–1922: Hospitals, Spectacle and Vice: The Making of Dublin City* (Dublin, 2006); Lindsey Earner-Byrne, *Mother and Child: Maternity and Child Welfare in Ireland, 1920s–1960s* (Manchester, 2007); Susannah Riordan, 'Venereal disease in the Irish Free State: The politics of public health', *Irish Historical Studies*, 35 (2007).

12 For some examples see Brendan D. Kelly, 'Poverty, crime and mental illness: Female forensic psychiatric committal in Ireland, 1910–1948', *Social History of Medicine*, 20:3 (2008), pp. 311–28; Leanne McCormick, '"The scarlet woman in person": The establishment of a Family Planning Service in Northern Ireland, 1950–1974', *Social History of Medicine*, 20:3 (2008), pp. 345–60; Lindsey Earner-Byrne, 'Managing motherhood: Negotiating a maternity service for Catholic mothers in Dublin, 1930–1954', *Social History of Medicine*, 19:2 (2006), pp. 261–77; Greta Jones, '"Strike out boldly for the prizes that are available to you", medical emigration from Ireland 1860–1905', *Medical History*, 54:1 (2010), pp. 55–74.

13 Greta Jones and Elizabeth Malcolm, 'Introduction: An anatomy of Irish medical history', in idem. (eds), *Medicine, Disease and the State in Ireland, 1650–1940*, pp. 1–2.

14 See for example Ann Daly, 'The *Dublin Medical Press* and medical authority in Ireland 1850–1890' (NUI Maynooth: unpublished PhD thesis, 2008); Caitriona Foley, 'The great flu epidemic in Ireland, 1918–19' (University College Dublin: unpublished PhD thesis, 2009).

15 MA in the Social and Cultural History of Medicine, School of History and Archives, University College Dublin.

16 The relevant websites are www.ucd.ie/history/body.htm and www.arts.ulster.ac.uk/history_medicine/index.htm

17 James Kelly and Fiona Clark (eds), *Ireland and Medicine in the Seventeenth and Eighteenth Centuries* (Farnham, Surrey, 2010).

18 Löwy, 'Social history of medicine: Beyond the local', p. 466.

19 For a comparative study see Mel Cousins, 'Poor relief and families in nineteenth-century Ireland and Italy', *History of the Family*, 13:4 (2008), pp. 340–9.

20 Deborah Brunton, *The Politics of Vaccination: Practice and Policy in England, Wales, Ireland, and Scotland* (Rochester, New York, 2008) and Laurence M. Geary, *Medicine and Charity in Ireland, 1718–1851* (Dublin, 2004).

21 Jones and Malcolm, 'Introduction: An anatomy of Irish medical history', pp. 4–5.

22 Catherine Cox, 'The medical marketplace and medical tradition in nineteenth-century Ireland', in Stuart McClean and Ronnie Moore (eds), *Folk Healing and Health Care Practices in Britain and Ireland: Stethoscopes, Wands and Crystals* (Oxford, 2010), pp. 55–79; Jones, '"Strike out boldly for the prizes that are available to you"'; James Kelly, 'Health for sale: Mountebanks, doctors, printers and the supply of medication in eighteenth-century Ireland', *Proceedings of the Royal Irish Academy*, 108C (2008), pp. 75–113; idem., 'Domestic medication and medical care in late early modern Ireland', in Kelly and Clark (eds), *Ireland and Medicine*, pp. 109–35. Margaret Ó hÓgartaigh and Margaret Preston (eds), *Gender, Medicine and the State in Ireland, Australia and the United States* (Syracuse, forthcoming).

23 Daly, 'The *Dublin Medical Press* and medical authority'; Foley, 'The great flu epidemic'; Susan Kelly, 'Suffer the little children: History of childhood tuberculosis in the north of Ireland, c. 1865 to 1965' (University of Ulster: unpublished PhD thesis, 2008); Elizabeth Lake, 'The history of medical communication' (University of Ulster, current doctoral candidate).

24 For example, the journal of the Irish Society for Archives devoted an issue to primary sources for medical history in Ireland see *Irish Archives*, 15:3 (2008). The National Archives of Ireland and the Royal College of Physicians in Ireland have been awarded funding grants by the Wellcome Trust to catalogue and preserve archives.

25 Some work has been completed on the history of Irish nursing and mid-wifery during the last decade. For example see Gerard M. Fealy, *Care to Remember: The Story of Nursing and Midwifery in Ireland* (Dublin, 2005); idem., *A History of Apprenticeship Nurse Training in Ireland* (London, 2005); Margaret Ó hÓgartaigh, 'Flower power and "mental grooviness": Nurses and midwives in Ireland in the early twentieth century', in Bernadette Whelan (ed.), *Women and Paid Work in Ireland, 1500–1930* (Dublin, 2000), pp. 133–47; Phil Gorey, 'A history of midwifery in Ireland' (University College Dublin, current doctoral candidate).

26 In 2006 Greta Jones, University of Ulster, was awarded funding to explore medical migration to and from Ireland from 1860 to 1960. Catherine Cox, University College, Dublin and Hilary Marland, University of Warwick, were awarded a project grant to examine the relationship between migration and mental health among Irish migrants to Lancashire, c.1850–1921. This project is currently underway.

27 Leanne McCormick, '"One Yank and They're Off": Interaction between U.S. troops and Northern Irish women, 1942–1945', *Journal of the History of Sexuality*, 15:2 (2006), pp. 228–257; and her chapter in this volume. Philip Howell, 'Venereal disease and the politics of prostitution in the Irish Free State', *Irish Historical Studies*, 33, 121 (May 2003), pp. 320–41; Riordan,

'Venereal Disease in the Irish Free State'. For work on the eighteenth and nineteenth century see Boyd, *Hospitals, Spectacle and Vice*; Geary, *Medicine and Charity in Ireland*; and Maria Luddy, *Prostitution and Irish Society, 1800–1940* (Cambridge, 2007).

28 See Brendan D. Kelly, 'Poverty, crime and mental illness: Female forensic psychiatric committal in Ireland, 1910–1948', *Social History of Medicine*, 21:2 (2008), pp. 311–28; E. Malcolm, '"Ireland's crowded madhouses": The institutional confinement of the insane in nineteenth- and twentieth-century Ireland', in Roy Porter and David Wright (eds), *The Confinement of the Insane: International Perspectives, 1800–1965* (Cambridge, 2003), pp. 315–33; E. Malcolm, '"A most miserable looking object". The Irish in English asylums, 1851–1901: Migration, poverty and prejudice', in John Belchem and Klaus Tenfelde (eds), *Irish and Polish Migration in Comparative Perspective* (Essen, 2003), pp. 115–26; Oonagh Walsh, *Land, Power and Politics: The Connaught District Lunatic Asylum and Irish Psychiatry, 1833–1910* (Syracuse, forthcoming 2010); idem., 'Gender and insanity in nineteenth-century Ireland', in J. Andrews and A. Digby (eds), *Sex and Seclusion, Class and Custody: Perspectives on Gender and Class in the History of British and Irish Psychiatry* (Amsterdam, 2004); idem., '"The designs of providence": Race, religion and Irish insanity', in J. Melling and B. Forsyth (eds), *Insanity and Society: The Asylum in its Social Context* (London, 1999); Catherine Cox, *Managing Insanity in Nineteenth-Century Ireland, 1832–1900* (Manchester, forthcoming 2010); Pauline Prior, *Madness and Murder: Gender, Crime and Mental Disorder in Nineteenth-Century Ireland* (Dublin, 2008).

29 Fiachra Byrne, 'Psychiatry and psychiatric care in 20th century Ireland', (University College Dublin, current doctoral candidate); Tom Feeney, IRCHSS Post-Doctoral Fellow, research project on 'The politics of mental health in independent Ireland: The making and unmaking of the Mental Treatment Act, 1945' at the Centre for the History of Medicine in Ireland, UCD.

30 For an exception see Mark Finanne, *Insanity and the Insane in Post-Famine Ireland* (London, 1981, 2nd edn, 2008).

31 For example see Terrence McDonough, *Was Ireland a Colony?: Economics, Politics and Culture in Nineteenth-Century Ireland* (Dublin, 2005); David Fitzpatrick, 'Ireland and empire', in Andrew Porter (ed.), *The Oxford History of the British Empire: The Nineteenth Century* (Oxford, 1999); Stephen Howe, *Ireland and Empire: Colonial Legacies in Irish History and Culture* (Oxford, 2000).

32 For example, the Centre for Medical History, University of Exeter, has a vibrant research strand in this field. See centres.exeter.ac.uk/medhist/about.shtml.

33 Keir Waddington, *The Bovine Scourge: Meat, Tuberculosis and Public Health, 1850–1914* (Woodbridge, 2006).

34 The workshop was funded by the Wellcome Trust Centre for the History of Medicine at University of Warwick.

35 See Ronald D. Cassell, *Medical Charities, Medical Politics: The Irish Dispensary System and the Poor Law, 1836–1872* (Woodbridge, 1997).

36 Luddy, *Prostitution and Irish Society*; Riordan, 'Venereal Disease in the Irish Free State'.

1
'Bleeding, vomiting and purging'[1]: The Medical Response to Ill-health in Eighteenth-century Ireland

James Kelly

Introduction

One of the main consequences of the primacy afforded the medical institution in the history of Irish medicine has been the failure to engage in any consistent way with what it meant to be ill, and with how the patient and the embryonic medical order responded to illness.[2] As a result, compared with the situation in England, the history of medicine in Ireland, in the pre-modern era particularly, is under-developed. Work underway will help to remedy this deficiency, but it is not possible *yet* to offer any secure assessment of public access to trained medical personnel, of the standard of medical assistance that was available, of the experience of illness, or of the impact on public health of the emergence of hospitals and dispensaries on the country as a whole in the late seventeenth and eighteenth centuries. It can be suggested, based on the dramatic expansion in the availability of medical information in print during the eighteenth century, that the Anglophone population of Ireland was better positioned than it had ever been to engage in a domestic setting in the diagnosis of illness and in the pursuit of programmes of care.[3] Moreover, the availability from the 1720s of a vastly expanded array of proprietary and patent medicine also permitted an increased resort to auto-diagnosing and to the treatment of illness within the domestic environment.[4] The full social implications of these developments for the treatment of illness have yet to be teased out, but the fact that they occurred against a backdrop of increasing consumption and an expansion in the number of regular and irregular medical practitioners such as also occurred in England suggests that the situation in Ireland bears comparison with that jurisdiction, where, Porter maintained, 'there was no single privileged medicine' because 'many types of medicine co-existed'.[5] This should not come as a surprise, given the shared linguistic, cultural, historical and religious horizons of the Anglophone populations of Britain and Ireland, but since the Anglophone community accounted for only 20 per cent of the population in Ireland, one cannot generalise from their situation for the country at large.[6] However, there is little to suggest that the largely Irish-speaking native population sustained a vibrant and effective culture of folk medicine. Indeed, the eclipse in the seventeenth century of the remaining Gaelic medical schools, which were anchored in the Galenist tradition, may well have left a vacuum that was inadequately filled by a serendipitous mixture of holy wells, faith healers, herbal potions, folk beliefs and magico-religious rituals.[7] It can also be suggested, based on the perceptible increase in the number of trained medical personnel across the island that the eighteenth century witnessed the extension of the diagnostic, commercial medical tradition that they – surgeons and apothecaries primarily – personified, and greater resort to their use. It cannot be said that the quality of the medical treatment provided improved greatly as a result, as the continuing ascendancy of the humoral conception of the body, deficiencies in medical regulation and oversight, and the limitations of medical education ensured that medical practitioners continued to offer inadequate responses to most conditions. However, practice was not unchanging and, while bleeding and purging remained fashionable longer than was advantageous or wise, these therapies were reinforced by the embrace of proprietary medicines in response to patient demand and complemented by more benign therapies such as 'drinking the waters'.

It is not possible in the space of a short paper to engage other than in a preliminary fashion with the many issues that the medical response to ill-health requires. Moreover, because this paper is confined to exploring the experience of those of the Anglophone population – middle and upper class – who have left an identifiable evidential footprint, it is necessarily incomplete socially. To compound these limitations, the failure to date to engage in a detailed analysis of the impact of the nascent hospital and dispensary system on the doctor-patient relationship, equivalent to that pursued by Mary Fissell for Bristol, largely precludes a wider conceptual engagement with the issue of the changing doctor-patient relationship.[8] As a result, if this paper is silent on the origins of the doctor-driven medical system that replaced the patient-driven system, it does (for the first time) seek to explore how patients and doctors responded to ill-health during an era when, for the middle and upper classes at least, the patient was in the ascendancy in the evolving doctor-patient relationship.

The ubiquity of pain and illness

For the population of seventeenth and eighteenth-century Ireland, illness was an omnipresent threat that could instantaneously plunge an individual into a maelstrom of debilitation and, if the condition proved intractable, lead the sufferer towards a painful and lingering death. This is vividly revealed at the societal level by the presence of such ostensibly curable conditions as colds, cholic, dropsy, piles, palsy, quinsy (tonsillitis) and bad teeth on the Dublin bills of mortality for the 1730s, as well as by passing references in private correspondence to the loss of 'vast numbers', as a consequence of 'epidemicall' disorders, which were rationalised as 'God's will'.[9]

Given this context, belief in the therapeutic value of medicine was an important, perhaps even a necessary, antidote to the recurrent anxiety of the population of Ireland that they might be subject at any moment to major contagion. The unease this prospect generated is exemplified by the decision of the bench of bishops in 1721, when the country apprehended a visitation of the plague then raging in the Mediterranean, to prepare a 'prayer' for use 'during the continuance of our danger'.[10] In fact, Ireland did not experience the plague then or at any other point in the eighteenth century, but the speed with which 'reports of its being already in this or the other distant part of the kingdom' circulated in 1721 warranted the bishops' intervention, and the regularity with which the Privy Council thereafter directed that shipping from suspect parts should be quarantined, attests to the perceived seriousness of the threat.[11] This was explicable given the prevalence of epidemic disease, and the ostensible frequency with which substantial numbers of people were, as Archbishop William King of Dublin observed of one such instance, seized by a 'violent disease that carrys off ... almost everybody'.[12] The early eighteenth century, certainly, witnessed recurrent epidemics of ague, fever (Dunkirk and red spot) and smallpox, and while the excess mortality varied greatly, reports to the effect that 'many dy' were imprinted on popular consciousness by the frequency with which they were linked to personal loss.[13]

The most persistent and consistently costly epidemic contagion in the early eighteenth century was smallpox, which may have accounted for 10 to 15 per cent of all deaths, 80 per cent of which were children under ten.[14] For those of a more resigned philosophical outlook and with greater experience of illness, such as Bishop Edward Synge of Elphin, the loss of a child to smallpox was not a source of surprise, but the emotional impact could be overwhelming in other instances. This was the case, for example, with Jenny Sampson, an acquaintance of Jane Bonnell, who had buried six of her 11 children by 1736.[15] The feelings of vulnerability engendered in contemporaries when faced with such stark loss of life were reinforced by those cases in which ostensibly minor infections escalated into life-threatening conditions. Thus in the case of the influential speaker of the House of Commons, William Conolly, it seemed to contemporaries in 1728 that the 'shock', induced by a 'cold which seized his limbs', caused him not only to lose the power of motion but also to require several months of attentive care before he was pronounced 'better'.[16] It is not surprising therefore that attentive parents such as Bishop Synge, for whom 'even [a] slight disorder occasions me ... the greatest anxiety', were unrelenting in cautioning their children 'not to run [any] risque' with their health. Synge was particularly concerned that his daughter Alicia, in her teens during the late 1740s and early 1750s, should avoid sore throats and colds, for though he recognised that they were not all equally 'dangerous', his conviction that 'a severe cold' acquired during menstruation 'laid the foundation of grievous irremediable disorders, which affect both body and mind', ensured that he maintained a fastidious watch on his daughter's health.[17]

It may appear that Bishop Synge's concern for his daughter bordered on the obsessive, but he was not untypical. Sir Laurence Parsons, for instance, advised his wife in 1805 that 'if any symptom of illness, however slight, should appear at any time in either of the little boys, you will immediately ask medical advice. For though they have been hitherto so healthy that I have little apprehension on this account, yet disorder makes such a rapid way, when ever it begins, in such little bodies, that often a momentary omission can never be repaired'.[18] If such expressions of concern seem alarmist in respect of persons who, to all outward appearances, were perfectly healthy, the point is that one could not presume on remaining healthy because even the most ordinary conditions could become life-threatening. This was made abundantly clear to Alderman Thomas Person of Dublin in 1719 when his wife experienced 'such violent shedding' as a result of a miscarriage that her family feared for her survival.[19] Catherine O'Brien brought her baby to term nine years earlier, but the physical demand of the birth and the emotional toll of the death soon afterwards of her daughter, from 'convulsions, which the doctors have said she brought into the world with her, and was occasioned by the great melancholy and troubles I lay under during all the time of my being with child', underline the dangers of pregnancy for women.[20] There was no equivalent for men, but an acute attack of cholic experienced in 1697 by William King, then Bishop of Derry, which caused him to fear for his life, ensured that he ever after regarded a period of 'more than ordinary health' as a blessing.[21]

King survived to the grand old age of 79, at least in part because he lived carefully, ate discriminatingly, and sought medical advice when indisposed; others were less prudent. Neither response was likely to make any difference when the illness or condition was life-threatening. Many among the elite contrived to reduce their exposure, as Anne, the daughter of Sir Samuel Cooke, the MP for Dublin, counselled, by keeping disease at a distance when practical. Thus when Captain Henry Caldwell fell ill with fever at Cooke's seat at Leixlip, in the summer of 1763, his relations Lady Elizabeth Caldwell and daughter departed promptly for Dublin 'for fear of the fever'. On another occasion, Anne and her mother-in-law, Mary Weldon interrupted a journey to Lady Jocelyn when they were advised that the measles were active at their host's place of residence.[22] Such precautions might seem well advised; the problem was that movement was as likely to contribute to the spread of contagion as to its containment. The army bore a particular responsibility in this respect, as in 1746 when the transfer of troops from Cork to Limerick and Galway was responsible for transmitting an 'ugly fever from which many ... died'.[23]

Death was the worst-case outcome of illness of course. Severe pain was a side effect from which relief was sought with no less urgency. The debilitating impact of the weeks of pain endured by Bishop Nicolson of Derry in 1719 and 1723, arising from a combination of toothache, kidney stones, gout and gravel, and the disabling 'great pain' experienced in 1762 by the under-secretary at Dublin Castle, Thomas Waite, from a persistent headache, served in both instances to prevent them from performing other than their most compelling duties. Bishop Nicolson, for example, took eight days (rather than the usual four) to make the journey from Dublin to Derry in September 1723, while Archbishop King had to be 'carried in and out of the churches' during his triennial visitation of the diocese of Dublin in 1724, because he was 'afflicted with the gout'.[24] King clearly found the experience undignified, but the embarrassment he felt was mild when compared with the impact of the racking pain experienced by Jonathan Swift in 1742 from an inflamed eye; it was so severe, 'five persons could scarce hold him for a week from tearing out his own eyes'.[25] This was a case, clearly, in which the limitations of contemporary medicine were all too apparent; in other cases, such as toothache, patients tried whatever remedies were available to them in an attempt to ameliorate the pain. These ranged greatly depending on social and geographical situation, and since there were many conditions in which relief was temporary if at all, it is hardly surprising that in instances of prolonged or acute pain, which were linked to a recognised serious illness, death was perceived as a welcome release.[26]

Medical assistance

Of course, only certain conditions were of such severity as to euphemise death; for a majority, the determination to overcome illness sustained a strong and appreciative demand for formal medical intervention. Ideally, the ill and the injured would have enjoyed ready access to the three main branches into which medical practitioners were formally divided: physicians, surgeons and apothecaries. The most powerful were the physicians, who diagnosed illness and recommended treatment regimens. Acutely status conscious and socially exclusive due to their position as university graduates and, in the case of the most eminent, as licentiates of the Royal College of Physicians, the most notable medical body in the kingdom, physicians personified the strengths and weaknesses of the existing system. Frequently *au fait* with advanced medical thinking and practice due to their attendance at some of the leading medical schools in western Europe, among which Leiden and Edinburgh featured prominently, they exemplified the hands-off approach that epitomised the gentleman practitioner. Indicatively, the Royal College of Physicians of Ireland admitted only 63 fellows, 51 candidates and 74 licentiates in the century between 1693 and 1793.[27] The number of functioning practitioners at any given time significantly exceeded what these figures suggest, as the presence of 'other physicians' on the published lists produced for the later eighteenth century attests, and these numbers were increasing, but were still too few to cater for more than a minority of the population. According to *The Medical Review* in 1774, there were only 61 physicians in the country. Five years later, the London-published *Medical Register* listed 41 physicians practising in Dublin; it listed 46 in 1783 and, while the number reached 77 in 1797, it was still insufficient for a city of 180,000 people.[28]

There were fewer physicians, to be sure, than there were surgeons, whose Dublin number rose during the same time from 55 in 1779 and 68 in 1783 to 100 in 1797.[29] As surgeons were trained on the job and specialised in the treatment of external diseases and in the performance of procedures that required surgical intervention, physicians treated them with condescension as mere practitioners of a manual craft. Their subordinate status was officially avowed when they were incorporated with the barbers, apothecaries and periwig makers in one guild in 1687. This proved entirely unsatisfactory; as a result, surgeons sought increasingly to distance themselves from guild membership, and their existence as a distinct and separate medical interest was acknowledged when they were empowered by law in 1784 to introduce a training and licensing system, comparable to that possessed by the Physicians, by the establishment of the Royal College of Surgeons in Ireland.[30] This was attributable in part to the more hands-on engagement with respect to illness that evolved in the course of the eighteenth century, epitomised by some surgeons' acquisition of advanced skills. Therefore, as well as their traditional functions such as dressing wounds, draining abscesses, setting bones, reducing fractures and dislocations, and tending venereal, eye and skin diseases, surgeons undertook increasingly complex operative procedures in the hospital system, to which an appreciating number of those who resided in Dublin were attached.[31] More generally, surgeon practitioners contributed to the enhancement of their standing with the public and their appeal to the ill by charging lower fees than physicians and, outside Dublin, where they were frequently the most skilled medical practitioners available, by tending to the medical needs of local communities.[32]

Apothecaries did likewise in their capacity as 'doctors of the poor'.[33] Though commonly perceived as the lowest rung on the medical ladder and charged with the task of making up and supplying medicines to physicians, in practice apothecaries performed a wide range of diagnostic and other functions, and were attended by patients from a wider social catchment than were physicians. Consequently, there were calls in the 1770s for apothecaries to be prohibited from 'attending families and prescribing as physicians'.[34] This was not, however, a practical possibility in small rural towns like Strokestown, county Roscommon, where the local apothecary was the only source of medical knowledge in an emergency, such as a coach accident in 1749, where, despite declining to prescribe physic, the local apothecary did trespass beyond his field of expertise when he 'blouded the foot' of a traveller.[35] This was not unusual, because the notional tripartite division in function between physicians, surgeons and apothecaries, which was more than blurred in large conurbations such as Dublin, had still less purchase in small towns. Be that as it may, the country's leading apothecaries were acutely conscious that the growing demand for their services, which encouraged an increase in the number of apothecaries in Dublin from 56 in 1779 to 87 in 1797, was dependent on their maintenance of high standards and on their ability to forge a particular identity.[36] They took a significant step in that direction in 1747 when they broke away from the surgeons, barbers and periwig-makers to form their own guild. But they made a still more notable advance four decades later when, in 1792, they founded an Apothecaries Hall 'for the preparation and sale of genuine medicines'. This paved the way for the inauguration of 'a committee of inspection', which was empowered to examine 'all shops and warehouses' where medicines were sold and to order the destruction of 'any thing pernicious or adulterated' that was detected.[37]

Such regulation was necessary because none of the recognised fields of medicine, or the emerging dental specialists, were able to police adequately their own members. Still less able to do so were the various healers, tooth-pullers, empirics and oculists, who comprised the irregular practitioners with whom recognised medics competed in the medical marketplace. It is noteworthy in this context that the Corporation of Apothecaries sought to encourage practitioners who were 'not free' of the guild to become part of the Apothecaries Hall in the 1790s, while the efforts by surgeons in the 1770s to 'discourage unqualified apprentices', as well as 'the intrusion of empyricks of every denomination', indicate that prior to the establishment of the College of Surgeons in 1784, irregular practitioners were an active presence in this sphere also.[38] The resulting medical free-for-all, in which the empirics and trained practitioners competed for business, might have been avoided if parliament had put an efficient regulatory regime in place, but it was not prepared to do so, or to assume the powers itself. As a consequence, the onus lay squarely with the customer to establish the bona fides of the practitioner to whom she or he turned for guidance.

This was inherently hazardous but, despite the limited prospect of a cure, it is apparent, based on the growth in the number of trained practitioners in Dublin from 151 in 1779 to 265 in 1797, that the public was increasingly disposed to look to those with expertise for medical guidance.[39] One needed deep pockets to obtain the best advice, for though there is no reason to believe, based on the situation in provincial England, that the middling orders did not have access to good medical knowledge, it is also apparent that the most able practitioners were beyond the resources of all but the elite. This observation is supported by the passing remarks of William Fitzwilliam, Lord Fitzwilliam's Irish agent, in 1757 that 'few people can afford to be sick', and in 1761 that he was 'in too poor circumstances to afford a family sickness of any large continuance'.[40] The financial situation of William Pearde of Bernardstown, county Cork, was still more precarious, which explains his gratitude to Lady Roche in 1737 for five guineas to assist with medical bills.[41]

The concerns expressed by Fitzwilliam and Pearde at their ability to meet their medical expenses may seem magnified, given that the standard 'ordinary fee' of a physician was one guinea for a consultation at his premises and two guineas if it involved a visit. However, these fees belied the true cost, as Bishop Synge demonstrated in 1750 when he authorised a payment of £100 to Dr George Daunt of Dublin, to whom he appealed for regular advice, as a sort of medical retainer.[42] In 1755 the payment of £4 17s. 6d. to the surgeon who treated Job Darling, a revenue officer based at Ennis who was severely cut on the face, indicates the order of surgical fees in cases of significant injury. Surgeons did not acquire the unenviable reputation for avarice that befell physicians, but the fact that the surgeon's bill (£11 7s. 6d.) for treating a patient in 1778 was more than ten times the sum (£1 2s. 8d.) paid the apothecary, offers an indication of the relative scale and order of costs.[43]

Despite the sums involved, and the substantial fortunes made by some physicians,[44] money was not the only determinant of who did, or did not, have access to medical advice. Geography was also an important consideration because of the uneven distribution of medical personnel. This problem was at its most acute in the late seventeenth and early eighteenth centuries when large parts of the county were simply bereft of medical expertise. Jonah Barrington's later recollection that because 'there was seldom more than one regular doctor in a circuit of twenty miles', people looked to the local farrier for medication provides a perspective onto the strategies rural communities were obliged to employ.[45] This naturally encouraged caution, and helps to explain Elizabeth Freke's advice to family members in the early years of the eighteenth century that they should flee to Ireland when indisposed and 'seek ... help in England'.[46] Interestingly, few outside the very elite seem to have done so, for though the shortage of practitioners in the countryside remained an abiding concern, there was still confidence that the most learned Irish doctors were, Robert Molesworth concluded in 1696, as good if not 'better than the English ones', particularly when it came to the care of infants and children.[47] The main problem was that the most able physicians, surgeons and apothecaries were concentrated in Dublin, with the result that it was commonly acknowledged that the metropolis was the preferred place to be when ill because, it was observed confidently in 1716, 'you have all human assistance for ... recovery'.[48] This greatly exaggerates the practical reality, but it is indicative that Bishop Synge declined to take his daughter Alicia with him on his annual summer visits to Elphin in the early 1750s because he did not trust the local medical practitioners. Given the abundance of reports and anecdotes exposing the incompetence and ignorance of country practitioners, Synge's caution was comprehensible. Still more notable, however, is the choice available to the population of Elphin between two surgeons (neither especially skilled) and a competent apothecary, since it illustrates that by mid-century the country was fast acquiring a network of medical practitioners.[49]

This trend helped to ameliorate the profound scepticism with which contemporaries reflexively regarded all branches of medicine. Jonathan Swift certainly felt this way, as he made clear in 1737 when he famously observed of the physicians he had consulted that 'I never received the least benefit for their advice or prescriptions'.[50] Others were encouraged to reach a comparable conclusion by what they observed. Archbishop King, for example, reported in 1715 that 'the surgeons' called upon to treat Marmaduke Coghill, the influential lawyer, for 'a swelling in the throat ... put him to too much pain, kept him long ill, and threw him into something like a fever'.[51] Ten years later, Judge Thomas Ward expressed more general reservation when he concluded that 'a good nurse keeper' was of more use than a doctor in caring for children with smallpox.[52] Guided by such knowledge, patients devoted a lot of time and effort to the identification of the medical practitioner who was right for them and, when they reached a decision, frequently contrived to maintain the relationship long term.[53]

Enduring relationships attest to the confidence that many patients invested in their doctors. Archbishop King is not atypical. He was attended 'for above thirty years' by Patrick Dun, the leading physician of the day, who assisted him in recovering from a chronic attack of cholic in 1697. Encouraged by this experience, during his lifetime King strongly counselled those of his acquaintance to seek guidance from medical specialists.[54] Others were equally keen to propagate such advice, and it is notable that Edward Synge was unrelenting in his instructions to his daughter to accede to 'the doctor's judicious care and attention'.[55] Such faith was easier to maintain in those physicians or surgeons who possessed a good

manner, diagnostic skill, a commanding presence and a sound reputation, which accounts in large part for the confidence vested in figures
like Patrick Dun and Edward Barry, the latter being one of the country's
most talented physicians in the quarter century prior to his death in
1776.[56] By contrast, practitioners who were unable 'to form a judgement'
as to the likely outcome of an illness, or who advised that an illness
should be allowed to run its course in the (frequently well-founded) belief
that the natural immune system would prevail, did not inspire confidence.[57] This notwithstanding, it is striking that patients chose implic-
itly to vest so much belief in the curative prowess of their doctors, with
the result that when illness prevailed more blame was attached to the
patient. This was especially true when patients did not follow the advice
of their physicians. Dean Ormsby of Derry is a case in point; he was
sharply criticised for refusing in 1699 to 'observe his physitians (sic)
orders' by, among other actions, choosing to 'let blood without advice',
following which ... he relapsed, fell into epileptick fitts and dyed'.[58]
Several decades later, Dr Thomas Kingsbury's refusal to 'follow the physicians prescriptions', 'his doctors' maintained, contributed to his demise.[59]
Death may well have been the logical outcome in both cases, but the
crucial point about these and other examples is that the blame was placed
on the shoulders of the patient and not the doctor.

To be sure, attentive physicians engendered confidence in patients
and the medical regimens they recommended by their actions. They
also generated confidence in medicine as a whole by demonstrating
that doctors did not see their patients simply as a source of income,
and were prepared, as happened when Richard Edwards of Dublin was
struck by illness in 1747, on occasions to go the extra mile and visit their
patients as often as 'twice a day for some time'.[60] This was also made clear
to the public, when, following the fatal injury caused to John Van Lewen,
a successful Dublin physician, as a result of a fall on a surgical knife in
1736, seven physicians and three surgeons gathered at his bedside in the
hope that their collective wisdom might be of help.[61] It was not, which
confirmed the opinions of those like William Drennan, himself a licentiate of the College of Physicians, who observed that when 'two physicians ... attend a patient' it was more likely to result in confused and
contradictory diagnosing that must hasten 'the sacrifice of the patient to
the doctors if not the disease'.[62] Drennan was not alone is his antipathy
to multiple consulting, but it remained commonplace among those who
could afford it because of the continuing lack of certainty as to what constituted reliable medical knowledge, and the limits of the care medical
practitioners proffered.

Treating the humoral body

In practice, of course, the weaknesses and limitations of the medical system were ultimately attributable to its limited understanding of human biology. According to traditional humoralist medicine, the body was regulated by four humours – phlegm, yellow bile, blood and black bile – and the well-being of the human body depended on maintaining the humours in a state of equilibrium. This was achieved in a healthy body through the processes of alimentation and evacuation, but in an unhealthy body, where the humours were imbalanced or 'acrimonious', well-being could only be restored by the controlled removal of the excesses that disrupted corporeal balance. If, according to this analysis, the role and function of the medical practitioner was straight forward, it was complicated by the reservations of some of the great doctors of the age, Frederick Hoffman of Halle and Hermann Boerhaave of Leiden, most notably, whose refinement, known as solidist theory, postulated that illness was a product of 'imbalances on the irritability of the solid fibrous components of blood vessels and nerves'.[63]

Since some of the most influential physicians practising in Ireland trained with Boerhaave,[64] they were familiar with his theories as well as his diagnostic methods and therapies, and, in respect of fever at least, embraced both the 'depletive' or 'evacuant' regimen and the stimulating or 'tonic' measures solidism advocated. In practice, as J. Worth Estes has observed, 'physicians easily assimilated solidist and chemical notions into their ancient humoral concepts'. They were assisted in this respect by the fact that the 'evacuant' regimen encouraged traditional strategies such as drawing blood, administering emetics and cathartics, and applying blisters, though for different reasons. The emphasis on diet, and the inclusion of water among the tonic drugs, was also easily accommodated, with the result that, though one can identify solidist influences, the response to illness practised in Ireland in the eighteenth century remained firmly humoralist.[65] It certainly provided the interpretative frame of reference of the Irish elite, demonstrated by the presence in correspondence of references to 'bad humours' and to the suggestion that one must 'be opened' to remove such troublesome humours.[66]

The potential of humoral medicine to be very physically demanding was demonstrated in 1686 when Sir John Perceval, the county Cork magnate, was struck down by fever. Treated initially with a posset and, when this did not work, with an emetic (a clyster), the effort to redress the humoral imbalance to which his illness was attributed commenced in earnest thereafter, with the withdrawal from alternate arms of eight ounces of blood. As this extracted blood was thought to be 'very corrupt and enflamed', the procedure was believed initially to have eased Sir John's temperature ('heat'). Perceval's relief was short lived, however, and following a further relapse, and the recruitment of a third physician, he was bombarded with a panoply of decoctions and powders before, in a last desperate effort, the physicians removed the blisters and plaster that had also been administered, and were contemplating cupping when Perceval died.[67] Forty years later, in October 1729, William Conolly, then terminally ill, was subjected to the regimen, only slightly less severe, of blisters on his arms and thighs, and to repeated purgings, which proved no more helpful. Indeed, the primary effect of this treatment was to discommode the final days of a man whose wish was to die 'in a composed manner'.[68]

These and other instances notwithstanding, 'bleeding, vomiting and purging' were, the state physician, Henry Cope, averred in 1737, 'the chief operations in physic by which diseases are cured and health preserved'.[69] Of the three procedures, bleeding was probably the most commonly implemented, though it was not without risk. It was reported from London in 1709 that the death of chief justice Pyne was 'hastened by bleeding two or three times', while in 1742, John Jephson, the archdeacon of Cloyne, 'died by having his arm cut off', as a result of an arterial puncture inflicted during a botched attempt to extract blood.[70] These were instances of medical misadventure; it was more usual simply for bleeding not to produce the anticipated improvement in health. When Anne Hamilton of Tollymore, county Down, was bled for an attack of cramp in 1710, it produced only an uncomfortably 'num[b]ing pain sometimes in one shoulder sometimes in another and a tingling or pricking through my veins and a weakness in my arms and legs'. Others observed that bleeding, employed in the hope that it would 'help my breadth and head', did 'little good'.[71] It may be that in instances such as these bleeding was an imprudent strategy. It certainly seems inappropriate to take blood from 'under the tongue and in the arme' from a woman who was labouring to breathe from an unspecified 'epidemic' illness that had laid her low for six weeks.[72] This conclusion is reinforced by suggestions that inappropriate amounts of blood were drawn, though Irish practitioners may have been more cautious in this respect than their English equivalents.[73] If so, it did not reflect any lack of belief in the efficacy of the procedure. Bishop Synge, for example, observed complacently in 1752 that the improvement in the condition of the wife of one of his clergy, who suffered from a 'cold', was 'owing to a second bleeding'. More generally, William Buchan, the great medical populist whose compendious text was afforded an honoured place in many Irish homes, advised that it was 'proper to bleed the patient' prior to the administration of a pectoral bolus for 'colds and coughs of long standing, asthmas, and beginning consumptions', while in the case of treatments for venereal disease, the *practice* was to draw 12 ounces 'in order to attenuate the blood, so that there might "be room for it after it is rarefied by the mercury"'.[74]

In addition to the overwhelming belief that medical benefits would accrue from this procedure, blooding was frequently performed in association with other treatments. In cases of patients with venereal disease, mercury was administered orally, by unction or by fumigation, once the body had been prepared by the drawing of blood. This was repudiated by some, including John Wesley, who recognised the dangers of mercury, but Buchan countered his advice, reflecting his general disposition to endorse, when he did not encourage, extreme medical intervention.[75] By comparison, following or accompanying bleeding with cupping or the application of blisters was non-controversial. Cupping of the skin in order to stimulate the humours was a well-established medical strategy. The physical bruising that resulted exaggerated the physical demand of the procedure, and the recommendation to the Irish under-secretary Thomas Waite in 1769 that he should try 'cupping' every four months, to ease the persistent pain he experienced in the head and shoulders, bears witness to the continuing belief in its efficacy as the end of the eighteenth century approached.[76]

Blistering, which solidists believed countered underlying inflammation, was employed still more frequently in response to a variety of conditions, though it too could be physically demanding when, as happened frequently, the blisters did not rise. Thomas Kingsbury, for example, was 'blistered from head to foot' without success, in a vain attempt to save his life in 1747.[77] A decade later, the wife of William Fitzwilliam, who was seized by convulsions shortly after giving birth, had blisters applied to her legs in advance of the arrival of her physician. Fitzwilliam's confidence in the efficacy of blisters was based on experience, since a combination of bleeding and a blister administered by Edward Barry in 1754 were credited with curing his infant son of inflammatory fever.[78] Edward Synge certainly placed great store on the procedure, for when a combination of bleeding and physic did not cure the wife of one of his clergy, he happily endorsed recourse to blistering and, convinced that it had cured the agronomist Charles Varley of a potentially life-threatening fever in 1750, he did so unreservedly.[79]

Synge's attitude tends certainly to affirm John Wesley's bemused observation in 1773 that in Ireland blisters were applied 'for anything and nothing'.[80]

Purgatives were only slightly less commonly used, though they too were not without risk, particularly when self-administered. Since it cost 2s. 6d. to secure a surgeon to draw blood, one can understand why Matthew Fountain, a shoemaker in Dublin, chose to auto-administer a tobacco clyster. However, he misjudged the risk and died in consequence.[81] This outcome was unusual as well as unfortunate but, since clysters served 'not only to evacuate the contents of the belly, but also to convey very active medicines [opiates notably] into the system', it was vital that they were used with discretion. They were employed to better effect to remove 'small worms lodged in the lower part of the alimentary canal', as well as in response to more serious conditions such as dysentery and bowel complaints.[82]

Purges performed a comparable function, though the fact that Sir Patrick Dun directed Archbishop King in 1709 to take a 'vomit and purge' in the full knowledge that 'they would signifie nothing to my chollick' suggests that they were not always therapeutically justified.[83] Patients evidently believed in their efficacy, however, based on the conviction that 'a pretty smart purging' was of 'great use'; Bishop Synge, for example, entertained no doubts as to their value. Laudanum was an active ingredient in some purges, but because they could be made 'mild' or 'strong' according to requirements, they appealed to patients and practitioners alike.[84]

Eighteenth-century patients acceded to these severe treatments not only because they were ordained by the humoral conception of the body but also because they *believed* they would assist in the alleviation of certain illnesses. Yet neither medical interpretations nor the patients' perceptions were static, and solidest theory, together with the recognition in seventeenth-century France and Great Britain that a regime of water drinking could have positive health consequences, encouraged an increasing recourse to safer therapies of which the use of spa waters was the most notable.[85] Water was already acknowledged as a valuable option in 1696 when the lord lieutenant, Earl Capell, pronounced that he had resolved, 'by advice of my physicians', to spend some time at Chapelizod, which was the spa closest to Dublin.[86] A decade earlier, William King, then bishop of Derry, spent a month drinking the waters in county Wexford but, in a significant shift in direction, he went to Bath for five months in 1697 to assist with his recovery from cholic, and he was so impressed that he subsequently drank Bath water in preference to beer.

Within a decade, water drinking had attained general social respectability, as Lady Dun made unambiguously clear when she observed in 1706 that she not only drank water for its health benefits, but also had joined 'a club' for that purpose.[87] This was a cheaper alternative than travelling to Bath, which was already acknowledged as the premier spa in the British Isles. Spa water certainly possessed great appeal, since as well as the large number who travelled to Bath or further afield in the course of the eighteenth century, many more consumed imported spa waters.[88] Moreover, as knowledge spread about the qualities of the waters at the various spas, native and foreign, 'drinking the waters' acquired a particular social, as well as medical, cachet. Thus Bishop Synge advised his daughter in 1747 that she should consume German Spa water (then available for purchase from the German Spa Water warehouse on Dame Street) with salts for best effect, and cautioned her against eating 'fruit ... or garden things', which the prevailing orthodoxy maintained was an inappropriate combination.[89]

Doctors too responded to the enthusiasm for water therapies by directing patients they believed would benefit to one or other of the major spas. Bath was the most popular destination for members of the elite but, as belief in the efficacy of balneotherapy increased, and as Irish spas were discovered, local alternatives were utilised.[90] No Irish spa ever came close to matching the appeal of Bath, but Mallow, Swanlinbar and Castleconnell were popular for a time, and the recognition of the medical efficacy of water stimulated interest in other natural products. One of the most enduring was whey, which was central to Patrick Dun's successful rehabilitation of Bishop King in 1697 and it acquired a particular, if short-lived, niche in Irish consciousness, in the form of goats' whey, which for a time brought members of the Irish gentry to south county Down to consume the product *in situ.*[91] Others were attracted by the therapeutic merits of salt-water bathing, buttermilk consumption, and by Bishop Berkeley's endorsement of tar water, which even secured a cautious commendation from William Buchan.[92]

As the limitations of a purely or predominantly medicinal approach were obvious and the value of what were essentially foodstuffs (whey and buttermilk) or natural products (water) was recognised, it was both logical and inevitable that doctors as well as patients should perceive the medical merits of a health regimen based on rest and recuperation and a diet of simple but wholesome food. On the face of it, this had little in common with humoralism, but there was a long tradition of writings, dating back at least to the mid-sixteenth century, which supported the view that a temperate lifestyle and a simple diet made the recipe for a long and healthy life. The physician, James Reynolds, made this point vividly in 1793 when he pronounced that 'pure air and exercise are indispensable for the preservation of my life'.[93] Others emphasised the merits of exercise, and the wisdom of abstemious eating, which echoed the advice forthcoming from the health guides written by physicians like George Cheyne.[94] The problem with such direction was that it required exceptional discipline, implicit trust in one's physician, and a willingness to engage in lengthy bed-rest to encourage the body's natural immune system when it was prescribed to combat illnesses. One individual who lived by such rules was Archbishop King. Following his brush with death in 1697, King was particularly attentive where his health was concerned, and he did not hesitate when he felt poorly to follow the guidance of his physician or to take to his bed, even if this necessitated his missing important business of state.[95] King did not avoid taking physic, but it is apparent that both he and the doctors who advised him perceived the value of permitting the body to recover naturally. They were not alone in recognising this, of course; others did likewise, doubtlessly encouraged by the realisation that while physic did not figure prominently on the extensive list of causes of mortality on the annual bills of mortality assembled during the first half of the eighteenth century, its presence there was another reminder of the hazards of medical care and the limited ability of medical practitioners to cure the humoral body.[96]

Conclusion

The limitations of the medical service available to the population of Ireland in the late early modern period were crucial factors in prompting contemporaries to assume direct responsibility not only for their health but also for their recovery when they fell ill. Since, for those who have left a record of their experiences, illness was routine, it is hardly surprising that people devoted so much time and effort to building up collections of medical receipts, and, when proprietary medicine became readily available in the 1720s, why it rapidly became one of the primary engines of a burgeoning consumer economy.[97] We know as yet too little of purchasing patterns generally, and still less of the amount of proprietary medicine purchased by individuals to comment with conviction on this development. Nevertheless, a diverse range of individual consumer decisions attest that the voluminous advertising devoted to popularising proprietary medicines was not without impact: for example, Bishop Nicolson of Derry expressed his confidence in the Balsam of capivi; Edmund Spencer of Renny, county Cork, asked a friend to send him 'a little of the Jesuit Drops'; Mrs Upton of county Down in 1772–3 purchased four boxes of James Pills and a dozen packets of James powders; and the Countess of Kildare swore by Ward's drops when her stomach was upset.[98] Indeed, it would appear, based on the preparedness of respected physicians like Patrick Dun to incorporate water into the therapies they pursued in the early eighteenth century, and of others later in the century to accommodate the use of James powders, that medical practitioners were so well aware of the limits of their humoral approach, that they were quite willing to embrace more modish non-medical remedies.[99] Their recourse to such commercialised remedies was perhaps appropriate because the eighteenth century witnessed few significant therapeutic breakthroughs, and those that did occur, such as inoculation, were not universally accepted. This is not to aver that access to medical assistance or medical practice was static. There were important structural developments symbolised by the foundation of hospitals and regulatory institutions such as the College of Surgeons and the Apothecaries Hall. Furthermore, the growth in the number of trained practitioners suggests that the public had greater recourse to their use, even if the public they treated was far from convinced of the efficacy of what they prescribed, and access was more problematic outside the main urban centres.[100] Given this backdrop, it is tempting to suggest that in Ireland as in England, trained medical personnel assumed the initiative in the care of illness in the eighteenth century, paving the way for the 'doctor-led medical economy'. However, the fact that most of the major institutions with which this process has been associated in England appeared later in Ireland, favours the conclusion that this was essentially a nineteenth-century development.[101] It was certainly the case that the moneyed classes, on whom this paper has focused, continued to possess enormous agency in the ways in which they dealt with ill-health. This was symbolised by the readiness with which some chose to look to England for assistance,[102] but it is equally striking that others concluded that accessing English medics conferred no advantage. Commenting on the physicians he encountered in Dublin in 1802, the editor of the *Edinburgh Magazine*, Robert Anderson, observed approvingly that they were 'not inferior to the same classes of men in Britain'.[103] This might be read as a back-handed compliment, but it does suggest that the developments that had taken place over a period of a century and a half, though offering only slight assurance of recuperation and little relief from pain, meant that more people had access to medical care and some reason to believe in the possibility of a medical cure to their ills and ailments.

Notes

1 Henry Cope, *Medicina Vindicata, or Reflections on Bleeding, Vomiting and Purging in the Beginning of Fevers, Smallpox, Pleurisies and Other Acute Diseases* (London, 1737).

2 For pertinent commentaries on the state of medical history in Ireland see Greta Jones and Elizabeth Malcolm, 'Introduction: An anatomy of Irish medical history', in idem. (eds), *Medicine, Disease and the State in Ireland, 1650–1940* (Cork, 1999), pp. 1–20; 'Introduction', in James Kelly and Fiona Clark (eds), *Ireland and Medicine in the Seventeenth and Eighteenth Centuries* (Farnham, Surrey, 2010), pp. 1–17.

3 James Kelly, 'Domestic medication and medical care in late early modern Ireland', in Kelly and Clark (eds), *Ireland and Medicine*, pp. 109–35.

4 James Kelly, 'Health for sale: Mountebanks, doctors, printers and the supply of medication in eighteenth-century Ireland', *Proceedings of the Royal Irish Academy*, 108C (2008), pp. 75–113; Roy Porter, *Health for Sale: Quackery in England 1660–1800* (Manchester, 1989); Irvine Loudon, 'Provincial medical practice in eighteenth-century Britain', *Medical History*, 29 (1985), pp. 25–6.

5 Roy Porter, 'The people's health in Georgian England', in Tim Harris (ed.), *Popular Culture in England* (London, 1995), p. 126.

6 See Toby Barnard, *A New Anatomy of Ireland: The Irish Protestants, 1649–1770* (London, 2003) for a penetrating analysis.

7 For the decline of the Gaelic tradition see Charlie Dillon, 'Medicinal practice and Gaelic Ireland', in James Kelly and Fiona Clark (eds), *Ireland and Medicine in the Seventeenth and Eighteenth Centuries* (Farnham, Surrey, 2010), pp. 39–52; for healers see *Faulkner's Dublin Journal*, 7 July 1730, 3 March 1739; Liam Swords, *A Hidden Church: The Diocese of Achonry, 1689–1818* (Dublin, 1997), pp. 253–4; and for holy wells see Caoimhin Ó Danachair, 'Holy wells of county Dublin', *Reportorium Novum*, 2:1 (1957–8).

8 Mary E. Fissell, *Patients, Power and the Poor in Eighteenth-Century Bristol* (Cambridge, 1991); review by Roy Porter, in *Journal of Social History*, 26 (1993), pp. 873–4. For the best modern account of this subject in Ireland, see Laurence M. Geary, *Medicine and Charity in Ireland, 1718–1851* (Dublin, 2004).

9 Bills of mortality, 1733–38, *Dublin Gazette*, 6 October 1733, 6 April 1734, 5 April 1735, 5 April 1736, 7 January 1737, 8 April 1738; King to Annesley, 3 April 1718 (Trinity College Dublin (henceforth TCD), King papers, MS 2535 f. 137); Knightly Chetwode to [ ], 13 August 1726 in 'The Chetwood letters', *Journal of the Kildare Archaeological Society*, 9 (1918–21), p. 104; Edmond Spencer to Francis Price, 23 May 1745 (National Library of Wales (henceforth NLW), Puleston papers, MS 3580 f. 31); Nicolson to Archbishop William Wake, 17 January 1720 (Gilbert Library, Wake papers, MS 27 f. 250).

10 Nicolson to Wake, 21 November 1721 (Gilbert Library, Wake papers, MS 27 ff 307–09); Cartaret to Grafton, 14 September, October 1721 (The National Archives (henceforth TNA), SP67/7 ff 87–8).

11   Nicolson to Wake, 24 January 1721 (Gilbert Library, Wake papers, MS 27 ff 279); National Archives of Ireland, Proclamations, passim; TNA, SP67/12 passim.

12   King to Annesley, 3 April 1718 (TCD, King papers, MS 2535 f. 137).

13   King to Beresford, 16 October 1712, King to Clogher, 12 May 1720 (TCD, King papers, MS 750/4 f/51, MS 750/6 f. 76); Knightly Chetwode to [ ], July 1726, 13 August 1726 in 'The Chetwood letters', p. 104; Spencer to Price, 23 May 1745 (NLW, Puleston papers, MS 3580 f. 31).

14   Peter Razzell, *The Conquest of Smallpox* (London, 2003); Marie-Louise Legg (ed.), *The Synge Letters: Bishop Edward Synge to his Daughter Alicia, Roscommon to Dublin 1746–52* (Dublin, 1996), pp. 112, 458.

15   Jane Bonnell to [ ], 25 May 1736 (National Library of Ireland (henceforth NLI), Smythe of Barbavilla papers, MS 41578/9).

16   Worth to Bonnell, 18 June, 7 September 1728 (NLI, Smythe of Barbavilla papers, MS 41580/27).

17   Legg (ed.), *The Synge Letters*, pp. 8–9, 19–20, 189, 203, 250, 252, 283–5, 395–6, 449–50, 459–60, 471.

18   Sir Laurence Parsons to Lady Parsons, 30 May 1805 (Birr Castle, Rosse papers, D/10/1).

19   Pearson to Bonnell, 16 July 1719 (NLI, Smythe of Barbavilla papers, MS 41580/24).

20   O'Brien to Bonnell, 24 June 1710 (NLI, Smythe of Barbavilla papers, MS 41580/22).

21   King to Southwell, 21 Dec 1697, King to Hartstonge, 2 June 1712 (TCD, King papers, MS 750/1 f. 147, MS 2532 f. 31).

22   'Diary of Anne Cooke', *Journal of the Kildare Archaeological Society*, 8 (1915–17), pp. 113, 216.

23   Spencer to Price, 15 December 1746 (NLW, Puleston papers, MS 3580 f. 54).

24   Lucy Waite to Wilmot, 23 November 1762 (Public Record Office of Northern Ireland (henceforth PRONI), Wilmot papers, T3019/4410); Nicolson to Wake, 19 January, 7 February, 6 August 1719, 4 April, 7 September 1723 (Gilbert Library, Wake papers, MS 27 ff 205, 207–08, 321, 323); King to Ferns, 29 September 1723 (TCD, King papers, MS 2537 f. 174).

25   Whiteway to Orrery, 22 November 1742 in Harold Williams (ed.), *The Correspondence of Jonathan Swift* (5 vols, Oxford, 1963–5), V, p. 207.

26   Barrymore to Price, 5 April 1748 (NLW, Puleston papers, MS 3577 f. 11); Legg (ed.), *Synge Letters*, pp. 386, 464; James Kelly, '"I was glad to be rid of it"': Dental medical practice in eighteenth-century Ireland', in M.H. Preston and M. Ó hOgartaigh (eds), *Gender, Medicine and the State in Ireland, Australia and the United States* (Syracuse, forthcoming).

27   J.D.H. Widdess, *A History of the Royal College of Physicians of Ireland, 1664–1963* (Edinburgh, 1963), p. 71.

28   Ibid., pp. 71, 102, 104, 108; *The Medical Register for the Year 1779* (London, 1779), pp. 179–80; *The Medical Register for the Year 1783* (London, 1779), pp. 162–5; *The Irish Court Register for 1797* (Dublin, 1797), pp. 270–3.

29   *The Medical Register for ... 1779*, p. 181; *The Medical Register for ... 1783*, pp. 1656; *The Irish Court Register for 1797*, pp. 273–5.

30 Widdess, *Physicians*, p. 108; C.A. Cameron, *History of the Royal College of Surgeons in Ireland* (Dublin, 1886), chapters 3, 5; Loudon, 'Provincial medical practice', pp. 4, 6–7, 8, 12–13.

31 For surgeons, and the hospitals to which they were attached, see *The Medical Register for ... 1779*, pp. 182–5; *The Medical Register for ... 1783*, pp. 167–9; *The Irish Court Register for 1797*, pp. 254–65.

32 *Dublin Gazette*, 17 May 1735; *Hibernian Journal*, 4 October 1775; William Fitzwilliam to Lord Fitzwilliam, 24 July, 7 September, 7, 29 December 1756 (National Archives, Ireland (henceforth NAI), Pembroke Estate papers, 97/46/1/2/7/85, 87, 91, 93).

33 The term is John Fleetwood's: see *The History of Medicine in Ireland* (2nd edn, Dublin, 1983), p. 92.

34 *Freeman's Journal*, 13 January 1774.

35 Legg (ed.), *Synge Letters*, pp. 159–60, 165.

36 *The Medical Register for ... 1779*, pp. 181–2; *The Irish Court Register for 1797*, pp. 277–9.

37 *Freeman's Journal*, 12 January 1790, 18 August 1791, 16 August 1792; *Hibernian Journal*, 19 February 1790, 19 August 1791.

38 *Hibernian Journal*, 16 February 1790; *Freeman's Journal*, 13 January 1774.

39 *The Medical Register for ... 1779*, pp. 178–85; *The Irish Court Register for 1797*, pp. 270–9.

40 Loudon, 'Provincial medical practice', pp. 9–10; William Fitzwilliam to Lord Fitzwilliam, 15 December 1757, 31 March 1761 (NAI, Pembroke Estate papers, 97/46/1/2/7/134).

41 Pearde to Price, 8 July 1737 (NLW, Puleston papers, MS 3576 f. 19).

42 Legg (ed.), *Synge Letters*, p. 223.

43 Widdess, *The College of Physicians*, p. 106; Petition of Job Darling, 19 March 1755 (TNA, CUST 1/55 ff 146–7); Secret Service fund cashbook, 1777–80 (NLI, Heron papers, MS 4134 ff 23, 31); *Freeman's Journal*, 20 October 1781.

44 See A.C. Elias, *The Memoirs of Laetitia Pilkington* (2 vols, Athens, Georgia, 1997), pp. 16, 66, 234; J.B. Lyons et al., *The Irresistible Rise of the Royal College of Surgeons* (Dublin [1984]), p. 8.

45 Jonah Barrington, *Personal Sketches of His Own Times* (3 vols, London, 1827–32), III, pp. 31–2.

46 'The diary of Elizabeth Freke', *Journal of the Cork Historical and Archaeological Society*, 2nd series, 17 (1911), p. 98.

47 Molesworth to Molesworth, 29 October 1696 in HMC, *Reports on Manuscripts in Various Collections, VIII: Clements Papers* (London, 1913), p. 218. See also John Dunton, *The Dublin Scuffle* (Dublin, 2000), pp. 241–2.

48 James to William Smyth, 10 December 1716 (NLI, Smythe of Barbavilla papers, MS 41582/2).

49 Legg (ed.), *Synge Letters*, pp. 165, 230, 332, 477; W.R. LeFanu, 'Swift's medical friends', in E. O'Brien (ed.), *Essays in Honour of J.D.H. Widdess* (Dublin, 1978), p. 55; M.L. Legg (ed.), *Diary of Nicholas Peacock, 1740–1* (Dublin, 2004), p. 102.

50 Swift to Pulteney, 7 March 1737 in Williams (ed.), *Correspondence of Jonathan Swift*, V, p. 7.

51 King to Dromore, 17 May 1715 (TCD, King papers, MS 2536 ff 282–3).

52   Cited in John Stevenson, *Two Centuries of Life in County Down* (Belfast, 1917), p. 462.

53   Legg (ed.), *Synge Letters*, pp. 380, 464, 469.

54   King to Addison, 2 November 1714, 3 May 1715 (TCD, King papers, MS 2536 f. 104, MS 750/4 ff 253–4); T.W. Belchem, *Memoir of Sir Patrick Dun* (Dublin, 1866).

55   Legg (ed.), *Synge Letters*, pp. 15, 26, 189, 271, 464, 481.

56   Legg (ed.), *Synge Letters*, p. 481; William Fitzwilliam to Lord Fitzwilliam, 6 May 1755 (NAI, Pembroke Estate papers, 97/46/1/2/7/57); Lifford to Townshend, 19 August 1769, [26 September 1770] (Lewis Walpole Library, Townshend papers).

57   Pearde to Price, 16 March 1740, 12 July, 30 August 1747 (NLW, Puleston papers, MS 3579 ff 27, 119, 121).

58   King to Southwell, 4, 18 February 1700 (TCD, King papers, MS 1489/1 ff 122, 127–30).

59   Edwards to Price, 9 April 1747 (NLW, Puleston papers, MS 3577 f. 29).

60   Edwards to Price, 7 January 1748 (NLW, Puleston papers, MS 3577 f. 31).

61   *Memoirs of Laetitia Pilkington*, pp. 74–6, 459–62.

62   Drennan to McTier, February 1783 in Jean Agnew (ed.), *The Drennan-McTier Letters* (3 vols, 1998–99), I, p. 85.

63   J. Worth Estes, 'The medical properties of food in the eighteenth century', *Journal of the History of Medicine and Allied Sciences*, 51 (1996), p. 129.

64   See F.W. Innes Smith, *English-Speaking Students of Medicine at the University of Leyden* (Edinburgh, 1932).

65   See Worth Estes, 'The medical properties of food in the eighteenth century', pp. 127–35.

66   Bishop of Dromore to King, 26 February 1715 (TCD, King papers, MS 1995–2008 f. 1586); Legg (ed.), *The Synge Letters*, p. 400; Stevenson, *Two Centuries of Life in County Down*, p. 461.

67   Upton to Southwell, 29 April 1686 in HMC, *Egmont MSS*, II (Dublin, 1909), pp. 183–5.

68   Pearson to Bonnell, 23 October 1729 (NLI, Smythe of Barbavilla papers, MS 41580/24).

69   As note 1.

70   King to King, 14 January 1709 (TCD, King papers, MS 2531 f. 130); *Dublin Gazette*, 5 June 1742. The newspaper report that relates Jephson's death describes him as archdeacon of Cork, but this is a misattribution: see William Maziere Brady, *Clerical and Parochial Records of Cork, Cloyne and Ross* (London, 1864).

71   Stevenson, *Two Centuries of Life in County Down*, p. 467; Abigail Watson to Rachel Carleton, 2 February 1749 (NLI, Watson-Carleton papers, MS 5928/2).

72   Pearde to Price, 24 February 1749 (NLW, Puleston papers, MS 3579 f. 148); Carleton to Watson, 27 May 1751 (NLI, Watson-Carleton papers, MS 5928/2); Legg (ed.), *Synge Letters*, p. 198.

73   'Mrs Elizabeth Freke, her diary', *Journal of the Cork Archaeological and Historical Society*, 2nd series, 17 (1911), pp. 94–5, 98, 106, 147; Stevenson, *Two Centuries of Life in County Down*, p. 459; Resdall to Lincoln, 22 December 1754 (Nottingham University Library, Newcastle of Clumber papers, NeC 3649).

74 Buchan, *Domestic Medicine*, appendix; Legg (ed.), *Synge Letters*, p. 399; J.B. Lyons, *The Story of Mercer's Hospital, 1734–1991* (Sandycove, 1991), p. 25.

75 Stevenson, *Two Centuries of Life in County Down*, p. 459; Buchan, *Domestic Medicine*, appendix.

76 Barry to Price, 23 May 1749 (NLW, Puleston papers, MS 3577 f. 5); Wilmot to Waite, 28 February 1769 (PRONI, Wilmot papers, T3019/5899).

77 Edwards to Price, 7 January 1748 (NLW, Puleston papers, MS 3577 f. 5).

78 William Fitzwilliam to Lord Fitzwilliam, 5, 12 February 1754, 1, 15 December 1757 (NAI, Pembroke Estate papers, 97/46/1/2/7/33, 34, 104, 105).

79 Legg (ed.), *Synge Letters*, pp. 165, 230, 396.

80 Stevenson, *Two Centuries of Life in County Down*, p. 459.

81 *Dublin Gazette*, 22 August 1741.

82 Buchan, *Domestic Medicine*, appendix.

83 King to Bolton, 16 April 1709 (TCD, King papers, MS 2531 f. 76).

84 Legg (ed.), *Synge Letters*, pp. 141, 322, 328, 459–60; Buchan, *Domestic Medicine*, appendix.

85 See L.W.B. Brockliss, 'The development of the spa in seventeenth-century France', in R. Porter (ed.), *The Medical History of Waters and Spas* (London, 1990), pp. 23–47; R.S. Neale, *Bath 1680–1850: A Social History* (London, 1981).

86 Capell to Shrewsbury, 3 May 1696 in HMC, *Buccleuch and Queensbury*, II, p. 328.

87 Dun to King, 19 August [1696], Lady Dun to King, 9 February 1706, King to Addison, 3 May 1715 (TCD, King papers, MS 1995–2008 ff 1195, 2292, MS 750/4 ff 253–4).

88 Barclay to King, 3 May 1701, King to Addison, 14 May 1715 (TCD, King papers, MS 1995–2008 f. 786, MS 750/4 f. 255); Stevenson, *Two Centuries of Life in County Down*, pp. 464, 468; Hoare to Bonnell, 27 September 1712, Hoare to Newport, 24 August 1723, Smyth to Naper, 1736/7 (NLI, Smythe of Barbavilla papers, MS 41580/12/29, MS 41581/16); Nicolson to Wake, 1 September 1720 (Gilbert Library, Wake papers, MS 27 f. 268); Mathew to Fitzwilliam, 4 November 1742, 27 November 1748 (NAI, Pembroke Estate papers, 97/46/1/2/4/31, 5/57); Legg (ed.), *The Synge Letters*, pp. 8–9, 189, 356, 456, 466.

89 Legg (ed.), *The Synge Letters*, pp. 62, 66, 186, 223; Estes, 'Medical properties of food', pp. 136–41.

90 See John Rutty, *A Methodical Synopsis of Mineral Waters* (London, 1757); Martin to Fitzwilliam, 25 October 1748, Fitzwilliam to Fitzwilliam, 7 February 1758 (NAI, Pembroke Estate papers, 97/46/1/2/5/55, 7/106).

91 King to Addison, 3 May 1715 (TCD, King papers, MS 750/4 ff 253–4); Stevenson, *Two Centuries of Life in County Down*, p. 461; Legg (ed.), *The Synge Letters*, pp. 8, 26; James Kelly (ed.), *The Letters of Chief Baron Edward Willes, 1757–62* (Aberystwyth, 1990), p. 40.

92 Legg (ed.), *The Synge Letters*, pp. 281, 301, 386; Fitzwilliam to Fitzwilliam, 22 May, 16 June 1753 (NAI, Pembroke Estate papers, 97/46/1/2/7/15, 18).

93 J. Kelly (ed.), *Proceedings of the House of Lords, 1771–1800* (3 vols, Dublin, 2008), ii, p. 430.

94 Fitzwilliam to Fitzwilliam, 7 February 1758 (NAI, Pembroke papers, 97/46/1/2/7/106); Dromore to King, 29 January 1715, Ward to King,

25 March 1728 (TCD, King papers, MS 1995–2008 ff 1574, 2178); Estes, 'Medical properties of food', pp. 142–3.

95   King to Annesley, 27 January 1708, King to Irvine, 26 April 1716, King to Annesley, 28 October 1721, King to Southwell, 29 December 1725 (TCD, King papers, MS 2531 f. 44; MS 2533 f. 216, MS 750/6 f. 18, MS 750/8 f. 71).

96   King to Southwell, 25 January 1708, King to Henry Temple, 27 November 1718, King to Southwell, 29 December 1725 (TCD, King papers, MS 2531 f. 42, MS 750/5 ff 69–70, MS 750/8 f. 71); Legg (ed.), *Synge Letters*, pp. 189, 365; *Dublin Gazette*, 6 April 1734, 5 April 1735, 8 April 1738.

97   See Kelly, 'Health for sale', pp. 82–90.

98   Nicolson to Wake, 1 September 1719 (Gilbert Library, Wake papers, MS 27 f. 236); Spencer to Price, 16 June 1747 (NLW, Puleston papers, MS 3580 f. 59); Stevenson, *Two Centuries of Life in County Down*, p. 469; Dorothy Porter and Roy Porter, *Patient's Progress: Doctors and Doctoring in Eighteenth-Century England* (Oxford, 1989), p. 48; Janet Todd, *Rebel Daughters* (London, 2003), p. 76.

99   Above; Legg (ed.), *Synge Letters*, pp. 400–1, 426–7.

100 Memoirs of Rev Horace Townshend in Rosemary ffolliott, *The Pooles of Mayfield* (Dublin, 1958), p. 252; Kelly (ed.), *Proceedings of the House of Lords*, 1771–1800 (3 vols, Dublin, 2008), II, p. 430.

101 Fissel, *Patients, Power and the Poor in Eighteenth-Century Bristol*, passim; review by Porter, in *Journal of Social History*, 26 (1993), pp. 873–4.

102 See, for example, Marchioness of Buckingham to W.W. Grenville, 7 June 1789 in HMC, *Fortescue*, I, p. 437.

103 Anderson to Percy, 6 November 1802 in W.E.K. Anderson (ed.), *The Correspondence of Thomas Percy and Robert Anderson* (New Haven, 1988), p. 105.

2

General Practice and Coroners' Practice: Medico-legal Work and the Irish Medical Profession, c. 1830–c. 1890

Michael J. Clark

Introduction

For those familiar only with the current arrangements for the conduct of coroners' post-mortem examinations in Ireland, the very idea of general practitioners playing any significant role in medico-legal investigations must appear highly anomalous. The *Review of the Coroner Service* by the Department of Justice, Equality and Law Reform's Working Group, published in 2000,[1] makes no mention of general practitioners, and assumes as a matter of course that all medico-legal post-mortems will be carried out either by qualified or trainee hospital pathologists or by the State Pathologist or his/her deputy, an assumption further reflected in the definition of a coroner's post-mortem put forward by the Working Group for incorporation into the proposed new Irish Coroners' Rules.[2] Admittedly, the 1962 Coroners' Act, which for the time being remains the principal legal authority for current Irish practice and whose text is printed in full as an Appendix to the Working Group's *Review*, stipulates only that coroners' post-mortems shall be carried out by one, or occasionally two, 'registered medical practitioner(s)'.[3] However, it effectively excludes most general practitioners and hospital doctors from routine medico-legal work by specifying that 'A post-mortem examination under this Act shall not be made by a registered medical practitioner who had attended the person in relation to whose death an inquest is to be or is being held within one month of the person's death'.[4] Yet for most of the nineteenth century, the prevailing view was, in the words of one writer to the *Dublin Medical Press* in July 1842, that 'it is the duty of the coroner to summon the medical man who was in attendance on the deceased, or who lives near the place where the accident or death occurred', to carry out any post-mortem that might be required and to attend and give evidence at the inquest.[5] Unless the deceased died in hospital, an asylum, workhouse or prison, the post-mortem was usually, though not invariably, expected to be performed by the general medical practitioner who had last attended the deceased. In Ireland, as in England and Wales, many nineteenth-century coroners' post-mortems were carried out by surgeon-apothecaries and general physicians practising in the community, rather than by hospital or forensic pathologists, and attendance at inquests, the presentation of medical evidence and, rather less often, the conduct of coroners' post-mortems were relatively common events in the medical life and work of what would now be termed the 'general practitioner'.[6] Moreover, from the early years of the nineteenth century, a small but growing number of general physicians or surgeons were actually elected or appointed to coronerships throughout Ireland, rather more in proportion to the number of vacancies than in England and Wales during the same period.

For some nineteenth-century medical observers and commentators, however, the implications of the relationship between general medical practice and coroners' practice extended far beyond the detailed practical arrangements for medico-legal work. In Ireland, as in England, medical reformers and opinion-formers such as Arthur Jacob (1790–1874), co-founder and long-serving editor of the *Dublin Medical Press*, and Thomas Wakley (1795–1862), editor of *The Lancet*, saw the issues surrounding medico-legal work for coroners, and medical witnessing generally, as strategic opportunities to raise the consciousness of the newly-emerging Irish medical profession, to promote their own reform agendas, and to enhance medicine's public image and profile. Accordingly, medical evidence at coroners' inquests was the subject of frequent editorial comment and correspondence in the pages of the *Lancet* and the *Dublin Medical Press* during the 1830s and 1840s.[7] However, in very many instances, Irish general practitioners found that in practice, medico-legal work for coroners was fraught with difficulty, poorly remunerated, and attended by a whole raft of personal, professional and even political conflicts and controversies. This chapter will examine some of the reasons why medico-legal work for coroners was such an important issue for the Irish medical profession during the nineteenth century and why it nonetheless proved such a thankless and unrewarding task for so many practitioners.

Between a rock and a hard place: Medical practitioners and the provision of medical evidence at coroners' inquests in Ireland, c. 1780–c. 1850

In 1793, in the first original English-language work on forensic medicine to be published anywhere in Britain or Ireland, the Dublin surgeon

William Dease (1752–1798) remarked that 'To attend an inquest is, to the surgeon, unprofitable, the loss of time highly inconvenient, and in the event may be productive of very disagreeable consequences'. For these reasons, he added, 'no reflection can justly fall on gentlemen of the [medical] profession, for endeavouring to avoid an attendance that must so materially interfere with their other duties'.[8] At the time Dease was writing, medical evidence was already becoming a relatively common feature of proceedings in coroners' inquests and criminal trials in London, and Dease's work indicates that similar developments were taking place in Ireland, or at least in Dublin.[9] Indeed, it seems likely that a rudimentary fee structure for medical evidence was already beginning to emerge at this period, and in the early nineteenth century, it is clear that fees and expenses of up to £3 were sometimes paid to medical witnesses attending and giving evidence in coroners' inquests and criminal trials.[10] However, such payments were made at the discretion, not only of coroners and trial judges, but of the grand jury and this was to prove a major stumbling-block to the development of more trustworthy and reliable arrangements for the provision of medical evidence in Irish coroners' inquests during the nineteenth century.

More generally, coroners' inquests and criminal trials could prove severe ordeals for even the most skilful and experienced medical practitioners called upon to give evidence or, less frequently, to carry out a post-mortem examination. A practitioner summoned to examine a body or attend an inquest often had to travel a considerable distance to do so, sometimes in the middle of the night or in bad weather, while, in the nature of such cases, the examination often had to be carried out in difficult, unpleasant and primitive conditions, with no proper facilities or skilled assistance on hand. The practitioner might also have to wait upon the coroner's pleasure for a considerable time, with no assurance that he would eventually be paid for his time and trouble. Once the inquest had opened, as well as possibly bearing a heavy burden of responsibility for the imputation of guilt or innocence at a time when penal sanctions were often ferocious, the medical practitioner might have to face searching, often ill-informed, sometimes impertinent, and occasionally malicious questioning and criticism. Even though it was rare for interested parties to be legally represented at inquests, an attending surgeon-apothecary or physician's testimony could still be subject to close and detailed scrutiny by the coroner or, more rarely, by members of the coroner's jury and, as there were at this time no strict rules of evidence or procedure in the coroner's court, this process could hold some unpleasant surprises for the inexperienced or unwary practitioner. He would have to translate medical terminology

into layman's language and try to persuade the jury to accept the limitations of medical evidence and, in the process, would have to reckon
with the lawyer or layman's simultaneous demand for certainty and
intolerance of the incompleteness, ambiguity, and lack of consensus
that characterised much medical evidence and opinion. Finally, the
inquest might be held in a blaze of publicity, and either the coroner or
the coroner's jury might take it upon themselves publicly to criticise and
even censure the practitioner for his treatment of the deceased or for the
shortcomings of his evidence.[11] An appearance as a medical witness at an
inquest or a criminal trial could greatly enhance a doctor's reputation and
standing, but it was much more likely to prove a humiliating and even
disastrous experience for the unwary or incompetent practitioner, while
often leaving him seriously out of pocket. Yet at the same time, and
notwithstanding Dease's remarks, the opportunity to attend and give
evidence at an inquest was one which many general practitioners simply
could not afford to decline, for basic reasons of livelihood. It is hardly surprising, then, that advice to practitioners on how to conduct themselves
when giving evidence at coroners' inquests should have featured prominently in the teaching of medical jurisprudence, or that medical reformers should have tried so hard to reform the coronership and improve the
conditions and remuneration for ordinary practitioners attending and
giving evidence at inquests.

Reforming nineteenth-century Irish coroners' practice: The historical background

Dating back to at least the mid-thirteenth century in Ireland, the coronership had been part of the Irish administrative and judicial scene,
at least in the former English Pale, for nearly 600 years by the 1830s.
Having at first been concerned mainly with securing the Crown's financial interests through the imposition of various fines and amercements
and the forfeiture of deodands,[12] and only incidentally with investigating causes of death, from the sixteenth century onwards, the coroner's
duties were largely confined to the investigation of the causes and
authorship of sudden, violent or unexplained deaths in the community, and to the commitment for trial of persons charged with homicide by inquest juries.[13] Irish coroners' practice had already begun
to diverge significantly from its English counterpart by the end of the
Middle Ages, and some of these discrepancies were still apparent in the
early nineteenth century. For instance, in Irish practice coroners' fees
and expenses were scrutinised and approved by the county grand jury,

not by the magistrates, and the medieval idea of an overlap or inter-changeability between the functions and authority of magistrates, sheriffs and coroners also survived, so that each might replace or stand in for the other on certain occasions.[14]

During the nineteenth century, a number of attempts were made to bring Irish coroners' practice more closely into line with English practice, but Irish practice nevertheless continued to differ in several important respects down to the end of the century. The 1876 Coroners (Dublin) Act and the 1881 Coroners (Ireland) Act specified for the first time that all future appointees must have either legal or medical qualifications, thereby acknowledging the special importance of medicine for the work of coroners. However, the 1887 Coroners Act, which was to provide the basis for English practice for the next half-century, did not apply to Ireland, so that until the passage of the Local Government (Ireland) Act of 1898, Irish county coroners continued to be elected, were not paid regular salaries, and had to have their inquest fees, expenses and payments to medical and lay witnesses approved by the county grand jury.[15]

Medico-legal expertise in contention: Dr Richard Egan, the Irish Medical Association and the 'Dublin System' for procuring medical evidence at coroners' inquests, c. 1850–1881

For much of the nineteenth century, albeit for very different reasons, the coronership and the inquest were to prove troublesome and highly prob-lematic issues for both the Irish administration and the Irish medical and legal professions. During the 1820s and 1830s, which saw two well-publicised cases of financial irregularities involving coroners in counties Dublin and Sligo, the Irish administration's chief concern was to prevent the office of coroner from falling into disrepute, by encouraging wealthier and more respectable citizens to stand for election as coroners.[16] For its part, the Irish medical profession was mainly concerned with securing proper payment for practitioners giving medical evidence at inquests and carrying out coroners' post-mortems and, to a lesser extent, with securing the election or appointment of more medically qualified coroners, of whom there were eight out of a total of 84 in Ireland in 1843. The Coroners (Ireland) Act of 1846 mandated coroners investigating cases of sudden, violent or unexplained death to summon 'any legally qualified medical practitioner, being at the time in … practice at or near the place where such death happened', to examine the body and give evidence at the subsequent inquest.[17] Clearly, the wording of the Act could be interpreted in several different ways, but it was generally assumed that the practitioner 'at or near the place' would normally be the deceased's last medical attendant, or else any other local medical practitioner, and in Ireland as in England, many general practitioners saw themselves as having a kind of prescriptive right or prior claim to carry out post-mortem examinations and give evidence at inquests on their recently deceased patients, and to receive the coroner's fees for these services.

In Ireland, however, there had long been accusations that some coroners, especially in the city and county of Dublin, were giving post-mortem work to their own medical friends, to the exclusion of hard-working and often impecunious local medical practitioners, and even that some medical men had struck bargains with coroners to carry out post-mortem examinations and give medical evidence at reduced rates in return for a guaranteed monopoly of medico-legal work in that coroner's jurisdiction. Thus according to Henry Croly, writing in the *Irish Medical Directory for 1843*, 'The provisions of the [Coroners (Ireland) and Medical Witnesses] Act[s] are constantly violated ... There are numerous instances where ... the medical witness summoned is neither the medical attendant of the deceased, nor of the locality in which the inquest is held, nor [even] the most eminent of the profession brought from a distance.'[18] Such complaints were commonplace from the early 1840s onwards, but only in the 1870s did the Irish Medical Association (hereafter the IMA) take deliberate action to bring matters to a head.

For some time prior to 1876, the two alternating Dublin city coroners had been in the habit of entrusting much routine medico-legal work in the city to Dr Richard Egan, a surgeon formerly of Clontarf but, since 1870, conveniently resident in Talbot Street, barely a stone's throw from the city mortuary, in return for a fixed annual fee, variously said to be £75 or £150. In March 1877, however, in response to numerous complaints from its members that Dr Egan was thereby engrossing all the inquest fees for medical evidence, depriving the general practitioner of part of his livelihood and violating at least the spirit, if not the letter, of the Medical Witnesses Acts, the IMA, in the person of Archibald Jacob (1837–1901), son of Arthur, proprietor of the *Dublin Medical Press and Circular* and at that time Chair of Council of the IMA, formally challenged the audit of the city coroners' accounts for fees paid to medical witnesses. In April, the IMA proceeded to apply for a *certiorari* writ in the Court of Queen's Bench 'for the purpose of obtaining the opinion of the Court with respect to a decision made by the Public Auditor [to] allow a fixed sum yearly out of the Corporation funds to one ... medical man [Dr Egan] for his evidence at inquests'.[19]

As he was effectively denied access to the columns of the *Dublin Medical Press*, Egan defended himself vigorously in the columns of the *British Medical Journal* (hereafter *BMJ*), stating that the arrangement with the City Council had in fact existed since 1857, and claiming that no objection had been made to it when Sir George Porter (1822–1895), a former President of the College of Surgeons and Sergeant-Surgeon to Queen Victoria in Ireland, had occupied the post.[20] Nevertheless, in November 1877 the Court ruled unanimously that Egan's arrangement with the city coroners was indeed contrary to law, although it declined to make the ruling absolute and stated that the Corporation and the coroners had entered into the arrangement in good faith for reasons of economy.[21]

Deploring the way in which 'The sole question on both sides [seems to have] resolved itself into a pecuniary one', the *BMJ* lamented in November 1877 that 'One of the most important bearings of this case – viz. the procural of reliable medico-legal evidence at inquests – seems to have been altogether lost sight of'.[22] Stating that 'It must be patent to all, that a person familiar with normal and pathological anatomy is, when procurable, the proper individual to give evidence deriv[ed] from a *post mortem* examination; while the previous medical attendant, if any, should naturally give whatever *ante mortem* evidence he … possesse[s] … as to cause of death', the *BMJ*, far from congratulating the Irish Medical Association on the success of its action, argued that 'The Dublin system … might possibly be made the means of introducing a much sounder one in these cases than [presently] exists [elsewhere]'.[23] In reply, Thomas Grimshaw, MD (1839–1900), then Chair of Council of the IMA, denied that the Association's action amounted to no more than 'a mere scramble for fees', but rather constituted 'an important and necessary step towards the reform of coroners' law'.[24] Insisting that the IMA was 'not opposed' to the employment of qualified medical experts as medical witnesses, he stated that the Association's main objects in bringing the action had, on the contrary, been to ensure that the best available medical evidence was obtained, that the existing provisions of the law with regard to medical evidence were respected, and to put a stop to 'the system pursued by certain coroners of making bargains with particular medical men to attend and give evidence at inquests, on terms arranged between [themselves]', which he described as 'manifestly improper' and 'a great scandal to coroners' law'.[25]

As well as being Chair of Council of the IMA in 1877, Grimshaw was also a member (and future President) of the Council of the King and Queen's College of Physicians in Ireland, as the future Royal College of Physicians of Ireland was still formally known at this date, and was soon to become Medical Registrar-General for Ireland.[26] Whatever his true opinions on the vexed question of who should give the medical evidence at coroners' inquests may have been in 1877, three years later, in April 1881, Grimshaw was to be instrumental in the adoption of a *Report* by the College's Parliamentary Committee on a proposed new Coroners Bill for Ireland which took a very different view of the matter.[27] The *Report* described the present system 'of employing the practitioner "at or near the place", as provided for by Act of Parliament' – essentially, that for which the IMA had contended in 1877 – as 'a thoroughly unsatisfactory method of conducting such inquiries'.[28] Far from supporting the idea that coroners should be medical men, the *Report* recommended that they should be solicitors or barristers, appointed and paid for by the Crown rather than elected by the freeholders.[29] The *Report* went on to advocate that every coroner should have 'a skilled Medical Expert attached to his Court', who should also be appointed by the Crown to act 'as [a medico-legal] Assessor' to the coroner, and that this 'Medical Jurist and Pathologist' should be paid partly by salary and partly by inquest fees. In an attempt to placate the otherwise dispossessed general practitioner, however, the *Report* also recommended that in all cases where a medical practitioner had attended the deceased shortly before death, he should be examined as a witness and paid expenses and fees totalling not more than three guineas *per diem*, unless it was alleged that the death in question had been wholly or partly the result of inappropriate medical care.[30] In essence, then, the Parliamentary Committee's *Report* recommended that in the interests of procuring the best possible medical evidence at inquests, the system formerly practised by the Dublin city coroners should generally be adopted but put on a more systematic and financially transparent basis.

The issues raised in the Irish debates – namely, the true nature and role of medico-legal expertise, the relations between general practitioners, hospital doctors and pathologists, and the division of labour between law and medicine in the medico-legal investigation of sudden, unexplained death – did not differ fundamentally from those of contemporary English medical debates over the reform of coroners' practice. However, the Irish debates went much farther, were more clearcut in terms of the positions taken, and had more definite practical consequences, in the shape of the IMA's successful *certiorari* action, while nothing like the Dublin city coroners' system for procuring medical evidence at inquests seems to have existed anywhere in England before the 1890s and 1900s. Although there was much debate in England between 1877 and 1884 as to the possible merits of adopting the Scottish system of medico-legal investigation of sudden or unexplained deaths,[31] and although, as we have seen, the Irish case of *Jacob vs. Finlay* attracted considerable attention in the English medical press, no one seems to have advocated adopting anything like the 'Dublin system' for obtaining and paying for medical evidence until the new London County Council's Public Control Committee began to interest itself in the conduct of inquests in London in the late 1890s.[32] Already in 1793, William Dease had urged that 'The business of attending inquests... be made a distinct appointment, or annexed to some of the present professional establishments [i.e., the hospitals and dispensaries]'.[33] In this respect Ireland in general, and Dublin in particular, appears to have been rather precocious but, once allowance is made for the very different administrative and political contexts, it is remarkable how closely the expertise-based system of coroners' pathology recommended by the College of Physicians in 1881 resembles the medico-legal system for investigating cause of death in Ireland today.[34]

Medical practitioners and the problem of fees for medico-legal work in nineteenth-century Ireland

For the leaders of the Irish medical profession, the question of which practitioners were most entitled or best qualified to carry out coroners' post-mortems and to give medical evidence at inquests went to the heart of the troubled relationship between general medical practice and coroners' practice. But for most Irish medical practitioners, the most important and most frequently disputed question was the payment of fees for medico-legal work. As a Belfast surgeon observed in a letter to the *BMJ* in July 1888, 'The subject of fees to medical witnesses is one of almost equal importance and interest to the expert, the consultant, and the general practitioner'.[35] In theory, the whole question of fees for medico-legal work should have been dealt with by the Medical Witnesses Acts of 1836 and the Coroners (Ireland) Act of 1846, which laid down recommended scales of fees for different kinds of medico-legal work: £1. 1s. (one guinea) for examining a body and giving evidence at a coroner's inquest; another £1. 1s. for conducting a post-mortem and giving evidence; and up to £5. 5s. for toxicological analyses. These acts also allowed coroners to pay medical and lay witnesses their legitimate travel and subsistence expenses.[36] In practice, however, the whole question of fees for attendance and giving medical evidence at inquests was much more complicated and contentious in Ireland than in England, largely due to the continued existence and powers of the grand jury.

Under the Irish Grand Jury Act of 1836 and the Coroners (Ireland) Act of 1846, county coroners were obliged to 'present' for their fees and expenses, including fees to medical witnesses, to the grand jury at the next assizes, but the maximum sum for which they could present at any assizes was restricted to £50, or £65 in the case of borough coroners, provided that they did not thereby receive more than £2 per inquest in fees, and this figure was further reduced to £1. 10s. in 1846.[37] The effect of this was, firstly, to dissuade county coroners in particular from holding more than a very few inquests per assizes and, secondly, to make it difficult if not impossible for coroners to pay doctors according to the recommended scales of fees and expenses for attending inquests and giving medical evidence without finding themselves out of pocket. Moreover, despite the *Dublin Medical Press*'s repeated insistence to the contrary, it was widely alleged, and certainly believed, by many coroners, magistrates and grand jurors, either that the Medical Witnesses Acts did not apply to Ireland, or that the recommended scales of fees were not mandatory, and that grand juries had the discretionary power to vary or refuse payment of fees presented for, if they considered an inquest to have been held 'unnecessarily' or the medical evidence to have been superfluous to requirements. According to the 1872 edition of the *Irish Medical Directory*, in spite of a judicial ruling to the contrary in 1862, Irish county coroners were still often 'sorely harassed and put to great expense by having their claims for Inquests, or the fees they advance to medical men, disallowed by the Grand Jury'.[38] In consequence, cases frequently arose in which coroners either summoned medical men to examine bodies and to attend subsequent inquests, but failed to honour promises made to have their fees and expenses approved by the grand jury at the next assize, or in which medical men were eventually paid only a fraction of what they considered themselves entitled to under the Medical Witnesses Acts. This was all the more likely to happen when, as was frequently the case, more than one medical man was involved in a coroner's inquiry. Arthur Guinness, a surgeon of Clontarf, county Dublin, described his experience in one such case in a long letter to the *Dublin Medical Press* in March 1842:

An unmarried female was taken up on suspicion of having drowned her infant. A special summons was sent to me to attend the inquest on the body of the infant, and having examined the woman, to state whether she had been delivered of a child or not. I exculpated the woman of the dreadful charge of murder, as I satisfied myself by a stethoscopic and vaginal examination, that she was with child. The coroner gave me an order [for payment], and after calling on the county treasurer two or three times, I was at length informed that the coroner had given an order to another surgeon, and therefore they could not pay me, as the rule was, that two surgeons could not be paid on the same inquest. If the coroner knew this, as of course he did, it is nothing short of monstrous treatment ... I ... received nothing for my time and trouble.[39]

Together with the somewhat ambiguous wording of the relevant statutes concerning the procurement of medical evidence for coroners' inquests, the rather brutal capping of the total sums allowed for coroners' inquests naturally favoured the development of what another angry surgeon called 'the system of jobbing carried on by the coroners of this county [sc.: county Dublin], and certain medical men',[40] whereby one or two medical men agreed to carry out post-mortems and give medical evidence for lump sums in return for a virtual monopoly of medico-legal work in the coroners' districts. But while this may have favoured the development of medico-legal expertise, it still further exacerbated the bad feeling between coroners, medical men and grand jurors over the question of general practitioners' entitlement to coroners' fees. Indeed, the whole question of fees for attendance and giving medical evidence at inquests and conducting coroners' post-mortems was to bedevil the relationship between general practice and coroners' practice throughout the nineteenth century. Despite repeated complaints in the Irish medical press and persistent lobbying by the IMA, both the level of fees for medical witnesses generally, and the uncertainty of payment of coroners' fees to general practitioners in particular, remained live issues down to the end of the century.

Medical coroners in nineteenth-century Ireland

Thus far, we have only considered the relationship between general medical practice and coroners' practice from the standpoint of the services and evidence provided by medical men *to* coroners. But another aspect of the relationship becomes apparent when the coroner was himself a medical man. Until the 1881 Coroners (Ireland) Act stipulated that candidates for election to the coronership must have either a legal or a medical qualification, or else be JPs of at least five years' standing, there were no formal professional qualifications for the office.[41] Rather, under the provisions of the 1846 Coroners (Ireland) Act, coroners were required to possess an estate of inheritance worth at least £50 per annum, or a freehold estate worth at least £100 per annum, and to reside in the district, or at least the county of the district, in which they held office.[42] When Thomas Wakley, Radical MP for Finsbury and founder-editor of *The Lancet*, became coroner for the West Middlesex district in 1839, it was almost unheard of for a medical man to hold the office in England but, according to the *Irish Medical Directory for 1843*, at that time there were already eight (out of 84) medically qualified coroners in Ireland, including one each in Dublin city and county, and four others in Ulster.[43] By 1872, this number had risen to 20 (out of 95), including four out of five in Antrim and Belfast, two out of two each in county Louth and county Kildare, two out of four in county Limerick, and again one each in Dublin city and county, rising again to 27 (out of 92) by 1891.[44] By 1897, there were only 89 coroners in Ireland, but 34 (38 per cent) of them were medical men.

On the face of it, this constituted a far stronger representation for the Irish than for the English medical profession in the ranks of the coronership but, in fact, the underlying position was quite different in the two countries. From the time that Wakley first stood for election to the Middlesex coronership in 1830, the rival claims of the medical and legal professions to the office of coroner were widely debated in the English medical and general press and, for a time, support for a medical coronership became a kind of litmus test of commitment to the cause of general medical reform.[45] In Ireland, however, because of the general poverty of the country and the very small pool from which suitable candidates for local office could be drawn, the chief concern of the Irish administration was not so much which profession was better suited to hold the office, as to find any member of the local gentry or professional classes prepared to assume the office of coroner, in order to prevent it falling into what the preamble to the 1822 Act to Regulate the Qualifications of Persons holding the Office of Coroner in Ireland described as 'low and Indigent hands'.[46] While a local medical man with a small estate might not be ideal for the purpose, at least he could generally be relied upon to discharge the duties of the office with some degree of independence and probity, in the absence of any suitable solicitor or barrister. As the anonymous (legal) author of the article 'The office of coroners' in the *Irish Quarterly Review* for April 1859 explained, although in general lawyers, and especially barristers, were better suited to perform the duties of the office, particularly in large towns such as Dublin and Belfast, their appointment to county coronerships would be very unwise, as 'there is no demand for the professional services of barristers in country districts [i.e., in most of Ireland]', and their appointment as county coroners would therefore necessitate the payment of very substantial salaries to compensate them for the loss of any chance of private practice, which in turn 'would create a furore among indignant country gentlemen and payers of county taxes'. Country doctors, on the other hand, would suffer no such loss; for this reason, he argued, the remuneration of county coroners should be increased to £100 per annum 'so as to induce the most respectable medical men of the county to compete for the office'.[47] No such increase was in fact made, and the majority of coroners in both town and country continued to be drawn from the ranks of solicitors rather than barristers. However, it is noticeable that with the exception of Dublin, Belfast and their immediate hinterlands (counties Antrim, Dublin, and Kildare), most of the coronerships held by medical men in nineteenth-century Ireland were in fact in mainly rural areas, while several large towns, notably Cork and Limerick, never had a medical coroner until the twentieth century. For both the Irish administration and for many in the higher ranks of the Irish medical and legal professions, the key issue concerning the coronership was not so much whether it should be a medical or a legal office but, rather, whether anyone of sufficient credibility and standing could be found among the local representatives of the Ascendancy to prevent the office from falling into the hands of persons whose judgement and loyalties could not be relied upon in politically sensitive cases.

Medical evidence in a time of political struggle: Medical, local and national politics in the Mitchelstown and Fermoy inquests, July–August 1888

That such considerations were not merely hypothetical was demonstrated by several highly controversial inquest cases during the second half of the nineteenth century. Many of the vexations encountered by Irish general practitioners with respect to coroners' fees and rights of attendance at inquests were also experienced to some extent by English practitioners but, in addition, the Irish medical practitioner called upon to give evidence at a coroner's inquest sometimes had to contend with direct political interference by Dublin Castle and, occasionally, also with political pressure from nationalist opinion. This was seen at its starkest in 1888, when the inquest into the death of John Mandeville of Clonkilla, near Mitchelstown, county Cork, became a political *cause célèbre* for Irish nationalism and the object of indignant protests by many members of the Irish medical profession.[48] Mandeville was a locally well-known and comparatively well-off Catholic farmer who had long been active in the land agitation, and had been arrested under the Crimes Act after a National League meeting in Mitchelstown in October 1887, a month after the deaths of three other activists during political disturbances in the town had led to a coroner's jury returning a verdict of wilful murder against the County Inspector of Police and five other members of the Royal Irish Constabulary.[49] Together with several other well-known nationalists, Mandeville had been imprisoned in Tullamore Gaol, county Offaly, in November and December 1887 and, although he resumed some of his political activity after his release, he was alleged to have been in poor health for some months before his death on 8 July 1888. The subsequent inquest before Coroner Richard Rice, which lasted from 17 to 26 July 1888, heard testimony from 26 witnesses, including eight medical men, three Justices of the Peace and a Home Rule MP, and turned into a highly charged piece of political theatre, in which Dublin Castle went to great lengths to try to suppress or discredit any evidence that seemed to suggest that the regime in Tullamore Gaol might have caused or hastened Mandeville's demise, and to browbeat and cajole the inquest jury into returning a verdict of death from natural causes. To this end, Counsel for the Prisons Board sought not merely to suggest that Mandeville's health had been in decline for some time before his arrest, but also to discredit the evidence of the three local medical men who had attended Mandeville during his last illness.

Much of the nine days' worth of evidence was taken up by examination and cross-examination of the five local doctors who had either visited Mandeville in prison or attended him during his last illness. The medical evidence centred mainly on the (hypothetical) effects of close confinement in winter and of the prison's dietary regime and punishment on the deceased's health, and on the possible complications arising from the medical treatment he underwent after returning home from prison up to the time of his death. Dr James Ridley, the medical officer for Tullamore Gaol, whose evidence might have been expected to play a crucial role in the proceedings, did not in fact testify, as he committed suicide two days after the inquest opened. Counsel for the Prisons Board maintained that Mandeville's health had already been undermined before his imprisonment, by a combination of underlying constitutional weakness and alleged recent intemperate habits, and brought in Dr James Barr of Liverpool, Medical Officer for HM Prison, Kirkdale, Lancashire, who had never seen or treated the deceased, to defend the prison dietary regime and to discredit the testimony of the local doctors who had attended Mandeville. By his own admission, Dr Barr gave Counsel for the Prisons Board detailed advice on how to cross-examine the medical witnesses appearing on behalf of the deceased's family, and he did not hesitate to suggest that it was the local doctors' inappropriate treatment of Mandeville during his last illness, and not any alleged ill-treatment or malnourishment during his imprisonment, which had been largely responsible for causing, or at least hastening, his death. Barr was one of three medical witnesses who were called upon by Counsel for the Prisons Board solely to give *opinions* on the deceased's medical treatment, without actually having been present at any time during his last illness or having any first-hand knowledge of his condition or treatment whilst in prison. No post-mortem had been performed or was apparently considered necessary in this case, yet the greater part of the evidence presented was medical and Counsel for both sides spent much of the inquest trying to expose the alleged flaws and omissions in the medical testimony presented by the other side.

In his summing-up to the jury, the coroner, a local solicitor, denounced Dr Barr's attempts to discredit the testimony and damage the professional reputations of the local practitioners who had attended Mandeville during his last illness. After more than eight days of evidence, the jury took less than 40 minutes to return a verdict accepting the local doctors' judgement that the deceased had died of 'diffuse cellular inflammation of the throat', attributing this to 'the brutal and unjustifiable treatment he [Mandeville] received in Tullamore Gaol',[50] a verdict which the Unionist Chief Secretary for Ireland and future Prime Minister Arthur Balfour was to dismiss as 'entirely unwarranted by the facts of the case'.[51] Following the Coroner's suggestion, the jury also strongly condemned 'the vile aspersions of Dr Barr on the Doctors who attended Mr. Mandeville in his last illness'.[52]

Several of the participants in this case, including three of the medical witnesses, were reunited less than four weeks later in Fermoy, county Cork, for the inquest into the death of Dr James Ridley of Tullamore.[53] Here, the inquest jury eventually found that Dr Ridley had taken his own life 'whilst labouring under temporary insanity produced by apprehension of the disclosures at the Mitchelstown inquest',[54] but not before Counsel for the Ridley family had attempted to place much of the blame for Ridley's death on allegations made by Dr George Moorhead of Tullamore, one of the medical witnesses for the Mandeville family, in a letter to the *Freeman's Journal*. Meanwhile, not to be outdone, Counsel for the Mandeville family urged the Ridley inquest jury to return an ignominious verdict of *felo de se* on the deceased.[55] The Ridley inquest jury also repeated the Mandeville inquest jury's condemnation of Dr Barr's spoiling action and, in a final twist, went so far as to imply that Barr, not Moorhead, bore some of the responsibility for his late colleague's suicidal state of mind.[56]

In this case, although the weight of the evidence appeared to absolve Ridley from any direct responsibility for Mandeville's death, it was evident that in the embittered local political climate that had prevailed since the events of September–October 1887, medical men on both sides of the political divide were likely to find themselves caught in the crossfire between local nationalists and the Irish administration and to be faced with threats and even actual physical violence, as well as attempts to destroy their professional reputations, whether by other local doctors or by agents or representatives of Dublin Castle. In both the Mandeville and Ridley inquests, the local medical practitioners were at least formally vindicated, but not before they had been subjected to humiliating ordeals in the witness box and, to a lesser extent, in the press, and had their professional conduct and moral characters repeatedly impugned not only by agents of the State, but even by some of their professional colleagues and neighbours.[57] Admittedly, the Mandeville and Ridley inquests were quite exceptional, even for Ireland, but it is difficult to imagine English doctors being subjected to political pressures of this kind or intensity in a late nineteenth-century inquest, even one following (for example) a politically sensitive mining or industrial accident or a Fenian attack in England.

Conclusion

The relationship between general medical practice and coroners' practice, so central to the agenda and aspirations of medical reformers in both England and Ireland during the 1830s and 1840s, remained a peculiarly difficult one throughout this period. However, by abolishing the local administrative and judicial functions of the county grand jury and by turning coroners into salaried professional employees of the new county councils, the Irish Local Government Act of 1898 effectively removed two of the most frequent causes of contention in this relationship.[58] Other background social and demographic changes, such as the gradual lessening of mortality from famine and epidemic diseases, the increasing concentration of the population in large towns relatively well-provided with hospital beds, and the long-term declines in both levels of recorded lethal violence and in the Irish inquest rate from the 1850s and 1860s onwards,[59] must also have played their part in statistically reducing the number of occasions in which general medical practitioners, as distinct from hospital and dispensary doctors, would find themselves called upon to attend inquests or perform coroners' post-mortems. The declining inquest rate must itself have favoured the increasing concentration of medico-legal work in the hands of a few pathologists based in large general hospitals with relatively good facilities for conducting post-mortem examinations and other investigations but, in the absence of any detailed work on the history of pathology in relation to forensic medicine in Ireland in the nineteenth and twentieth centuries, this must remain largely a matter of conjecture. In particular, we lack any clear idea of the overall pattern and chronology of change in the social and professional organisation of medico-legal practice in twentieth-century Ireland. The contemporary systems for medico-legal investigation in both the Republic of Ireland and Northern Ireland are highly regulated, highly centralised and highly specialised, leaving little or no place for the general medical practitioner; but without a great deal of new research, it is impossible to say precisely how or when these systems emerged from the very imperfectly regulated, very decentralised and largely unspecialised medico-legal practice characteristic of nineteenth-century Ireland.

Notes

1 Department of Justice, Equality and Law Reform, *Review of the Coroner Service. Report of the Working Party* (Dublin, 2000).

2 Ibid., Appendix J, Part 1, 'Defination (*sic*) of terms', p. 156. See also Appendix J, Part 3, 'Post-mortem examinations', p. 157, and Part 4, 'Post-mortem by State Pathologist', p. 158.

3 Ibid., Appendix E, 'Coroners Act 1962', Clause 52-(1)(a), p. 139. In April 2007, the Irish Government published a new Coroners Bill, sponsored by Senator Mary O'Rourke, which is intended to replace the 1962 Coroners Act and implement a comprehensive reform of Irish coroners' practice. However, the relevant clauses of the new Coroners Bill (No. 33 of 2007) do not, in fact, significantly amend the existing provisions with regard to medical practitioners and the conduct of coroners' post-mortems. See Coroners Bill 2007, Part 10, Sections 74 (1), 75 (1) & 76 (6), at http://www.justice.ie/en/JELR/Pages/Coroners_Bill_2007

4 Ibid., Clause 52-(2)(a).

5 'Medical evidence at coroners' inquests', letter to the editor from Michael Doyle, surgeon, of Irishtown, county Dublin, *Dublin Medical Press*, 15 July 1842, p. 76.

6 For the position in England and Wales, see especially Ian A. Burney, *Bodies of Evidence: Medicine and the Politics of the English Inquest, 1830–1926* (Baltimore, 2000), Ch. 4.

7 For Wakley's interest in the Coronership and medical witnessing, see especially Elizabeth Cawthon, 'Thomas Wakley and the medical coronership:

Occupational death and the judicial process', *Medical History*, 30 (1986), pp. 191–202, and Burney, *Bodies of Evidence*, especially Ch. 1 and pp. 107–10. So far as I am aware, no serious study exists of Arthur Jacob as a medical reformer and publicist.

8   William Dease, *Remarks on Medical Jurisprudence, Intended for the General Information of Juries and Young Surgeons* (Dublin, 1793), pp. 29, 31. Dease described himself as 'Surgeon to the United hospitals of St. Nicholas and St. Catherine [Dublin]'.

9   For the early history of medical evidence in coroners' inquests and criminal trials in London, see especially Thomas R. Forbes, 'Crowner's quest', *Transactions of the American Philosophical Society*, 68 (1978), pp. 1–52; idem., *Surgeons at the Bailey: English Forensic Medicine to 1878* (New Haven and London, 1985).

10   Dease mentions an approved tariff for medical reports for magistrates of £2. 2s: Dease, *Remarks on Medical Jurisprudence*, p. 29. The Irish Grand Jury Act of 1836 mentions fees of up to £3 for attendance at inquests.

11   The printed transcripts of the Mandeville and Ridley inquests held at Mitchelstown and Fermoy in 1888 illustrate many of the potential pitfalls of medical witnessing at coroners' inquests. See below, notes 48 and 53.

12   A deodand – literally, something 'given to God' – was an object, whether animate or inanimate, deemed to have caused the death of the subject of a coroner's inquest. In such cases, a sum equivalent to the value of the object as determined by the inquest jury was declared forfeit to the Crown. Deodands were abolished in 1846.

13   For the early history of the Irish Coronership, see especially William G. Huband, *A Practical Treatise on the Law Relating to the Grand Jury in Criminal Cases, the Coroner's Jury and the Petty Jury in Ireland* (Dublin and London, 1896), pp. 1–30; A.J. Otway Ruthven, *A History of Medieval Ireland* (2nd edn, London, 1980), pp. 179–80; John L. Leckey and Desmond Greer, *Coroners' Law and Practice in Northern Ireland* (Belfast, 1998), pp. 1–6; Brian Farrell, *Coroners: Practice and Procedure* (Dublin, 2000), Ch. 1.

14   For these divergences between Irish and English practice and procedure, see especially Neal Garnham, *The Courts, Crime and the Criminal Law in Ireland 1692–1760* (Blackrock, 1996), pp. 97–8; Farrell, *Coroners: Practice and Procedure*, pp. 12–18.

15   Leckey and Greer, *Coroners' Law and Practice*, pp. 13–14; Farrell, *Coroners: Practice and Procedure*, pp. 27–32.

16   Leckey and Greer, *Coroners' Law and Practice*, p. 8; Farrell, *Coroners: Practice and Procedure*, pp. 20–3. For criminal cases involving coroners, see Henry Croly (ed.), *The Irish Medical Directory for 1843* ... (Dublin, 1843), p. 186.

17   9 & 10 Vic., c. 37, Clause XXXIII.

18   Croly (ed.), *Irish Medical Directory for 1843*, p. 311. The English Medical Witnesses Act of 1836 (6 & 7 Will. IV, c. 89), amended in 1842, allowed, but did not oblige, coroners to summon any qualified local medical practitioner to examine dead bodies and present their findings at inquests, and prescribed a scale of fees for making reports and giving medical evidence.

19   *British Medical Journal* [hereafter *BMJ*], 24 November 1877, pp. 740–1. A *certiorari* writ (Lat., 'to be informed of'), now known as a quashing order, was a common law writ issued by a superior court ordering a judicial review of a previous decision made by a lower court. For Egan, who was a Licentiate

of the Royal College of Surgeons in Ireland (LRCSI), see his obituary, in *Dublin Medical Press*, 9 November 1887, p. 459.

20 See Egan's letters to the editor, in *BMJ*, 24 March 1877, p. 375; 31 March 1877, p. 411; 14 April 1877, p. 472. Archibald Jacob wrote an indignant rejoinder to the first of these letters, see *BMJ*, 31 March 1877, p. 411.

21 For the court's decision, see *BMJ* (24 November 1877), p. 741.

22 Ibid.

23 Ibid.

24 'Coroners' medical witnesses and the Irish Medical Association', letter of Thomas Grimshaw to the editor, *BMJ*, 8 December 1877, pp. 829–30.

25 Ibid.

26 For Grimshaw, see especially his obituary, in *BMJ*, 3 February 1900, pp. 289–90.

27 Royal College of Physicians of Ireland, *King and Queen's College of Physicians in Ireland, College Journal*, Minutes of Council, 1 April 1881, p. 332. The *Report* of the Parliamentary Committee, dated 2 April 1881, is inserted between p. 345 and p. 346 of the *Journal*.

28 *Report* of the Parliamentary Committee of the King and Queen's College of Physicians in Ireland on the proposed Coroners (Ireland) Bill, 2 April 1881, p. 2.

29 Ibid.

30 Ibid., pp. 1–2.

31 See, for example, Douglas Maclagan, 'Forensic medicine from a Scotch point of view', *BMJ*, 17 August 1878, pp. 233–9, and subsequent editorial comments, in *BMJ*, 24 August 1878, p. 289.

32 For the London County Council's interest in coroners' pathology and inquests in the 1880s and 1900s, see David Zuck, 'Mr. Troutbeck as the surgeon's friend: The coroner and the doctors', *Medical History*, 39 (1995), pp. 259–87. See also Burney, *Bodies of Evidence*, pp. 123–4.

33 Dease, *Remarks on Medical Jurisprudence*, p. 31.

34 For the current Irish system of medico-legal investigation, see especially Farrell, *Coroners: Practice and Procedure*, and the Department of Justice, Equality and Law Reform Working Group's *Review of the Coroner Service* (see Note 1, above).

35 J. Dysart McCaw, FRCS, letter to the editor, *BMJ*, 28 July 1888, p. 207.

36 9 & 10 Vic., c. 37, Clauses XXVIII, XXXI and Schedule C.

37 6 & 7 Will. IV, c. 69; 9 & 10 Vic., c. 37, Clause XXV. See also Croly (ed.), *Irish Medical Directory for 1843*, p. 186, and Leckey and Greer, *Coroners' Law and Practice*, p. 8, n. 38, and p. 10, n. 50.

38 *The Irish Medical Directory for the Year 1872 ...* (Dublin, 1872), p. 153. This statement was repeated verbatim in the next (1891) edition of the *Directory*, on p. 321.

39 *Dublin Medical Press*, 7 (1842), p. 185.

40 Michael Doyle, surgeon of Irishtown, county Dublin, letter of 15 July 1842 to the editor, *Dublin Medical Press*, 8 (1842), p. 76.

41 44 & 45 Vic., c. 35, Clause II. See also Huband, *Practical Treatise on Law*, p. 1072; Leckey and Greer, *Coroners' Law and Practice*, p. 13; and Farrell, *Coroners: Practice and Procedure*, pp. 28–9.

42 9 & 10 Vic., c. 37, Clauses XVI and XXI.

43   Croly (ed.), *Irish Medical Directory for 1843*, pp. 188–90.
44   *The Irish Medical Directory for the Year 1872 ...*, pp. 154–5; *The Irish Medical Directory for the Year 1891 ...* (Dublin, 1891), pp. 322–3; *The Irish Medical Directory for the Year 1897 ...* (Dublin, 1897), p. 323.
45   See especially Burney, *Bodies of Evidence*, pp. 53–60, 80–2.
46   3 Geo. IV, c. 115, Preamble, quoted in Leckey and Greer, *Coroners' Law and Practice*, p. 8, and in Farrell, *Coroners: Practice and Procedure*, p. 20, n. 43.
47   Anon., 'The office of coroners – its practice and duties', *Irish Quarterly Review*, 9:1 (April 1859), pp. 268–82.
48   The following account is based largely on the 116-page transcript of the shorthand notes of the proceedings ordered to be published by the House of Commons in August 1888: *Mandeville Inquest ... Copy of Transcript of Shorthand Writers' Notes of Proceedings (Mr. Arthur Balfour). Ordered by the House of Commons to be Printed, 13 August 1888*, in *Balfour Papers 13/2. Ireland 1888: Miscellaneous Documents*, British Library.
49   For this earlier incident, see *The Mitchelstown Inquisition*, 2 L.R. Ir. 279 (1888), and discussion in Huband, *Practical Treatise on the Law*, pp. 255, 277–8. For background information see Laurence M. Geary, 'John Mandeville and the Irish Crimes Act of 1887', *Irish Historical Studies*, 25:100 (1987), pp. 358–75.
50   For Coroner Rice's summing-up, see *Mandeville Inquest ...*, pp. 108–14. For the inquest jury's verdict, see Ibid., p. 116.
51   For Balfour's comment, made during a Parliamentary answer to a question on the case on 26 July 1888, see *Lancet*, 4 August 1888, p. 238.
52   *Mandeville Inquest ...*, p. 116.
53   The following account is based largely on *Address of Mr. Ronan, B.L.* [counsel for the family of the deceased] *to the Jury at the Ridley Inquest, Fermoy, 17 August 1888*, in *Balfour Papers 13/8. Ireland 1888: Miscellaneous Documents*, British Library, supplemented by some articles in the medical press.
54   *BMJ*, 25 August 1888, p. 447.
55   *Address of Mr. Ronan ...*, pp. 2–4.
56   For the verdict, see *BMJ*, 25 August 1888, p. 447.
57   The highly politicised atmosphere surrounding the medical evidence in the Mandeville and Ridley inquests may be compared with that which attended the trial of Michael Cleary a few years later for the burning to death of his wife Bridget near Ballyvadlea, Co. Tipperary, in March 1895. See Angela Bourke, *The Burning of Bridget Cleary: A True Story* (London, 1999).
58   61 & 62 Vic., c. 37. For the effects of the 1898 Local Government Act, see especially Huband, *Practical Treatise on the Law*, 'Preface' to the 1911 Supplement, p. v, and Farrell, *Coroners: Practice and Procedure*, p. 32.
59   For the decline in recorded rates of lethal violence in Ireland from c. 1860 onwards, see Mark Finnane, 'A decline in violence in Ireland? Crime, policing and social relations, 1860–1914', *Crime, History and Society*, 1 (1997), pp. 51–70; and Ian O'Donnell, 'Unlawful killing: Past and present', *Irish Jurist*, 37 (n.s.) (2002), pp. 56–90, especially pp. 60–2. For the decline in the Irish inquest rate during the same period, see Leckey and Greer, *Coroners' Law and Practice*, pp. 10–11, and the corresponding n. 57.

3

Access and Engagement: The Medical Dispensary Service in Post-Famine Ireland

*Catherine Cox**

Introduction

Historians of medicine and the marketplace in early modern England have emphasised 'how little is known about health-care in rural areas'. As Ian Mortimer points out 'the idea of the "medical marketplace"', which has been central to the conceptualisation of medical provision and access, 'was developed for London and Florence'. Inherent in almost all discussions of the 'medical marketplace' is the assumption that urban-based practitioners ministered to their immediate neighbourhood, leaving the 'rural hinterland' reliant on heterodox practitioners.[1] Although the concept of the medical marketplace has proved particularly attractive in the study of the early modern period, understanding of medical provision and patients' behaviour in rural areas is patchy at best. Furthermore, in studies of the nineteenth century, historians have regarded the operation of the medical marketplace as less significant due to increased state involvement in medical provision, the consolidation of the medical profession and developments in scientific medicine. This has been particularly evident in work on nineteenth-century Ireland. The relatively precocious nature of state intervention in Ireland, originating in the late eighteenth century, has led some historians to assert that by the late nineteenth century, Ireland could boast 'one of the most advanced health services in Europe'.[2] The emergence and consolidation of an extensive government funded network of hospitals, infirmaries, lunatic asylums, dispensaries and workhouses during the nineteenth century would suggest that in a predominantly poor country with limited industrialisation, the dynamics of the marketplace were less influential.[3] It has been acknowledged that the provision of these services across the country was not uniform and certain regions, particularly poorer districts, encountered some difficulties in supporting medical and welfare institutions and personnel. As a consequence, historians have tended to assume that patients' continued use of heterodox practitioners remained popular due to the paucity of regular provision in certain regions, rather than as a result of patient choice.[4]

Central to the development of a medical system for the poor in nineteenth-century Ireland was the dispensary system, brought under the aegis of the Irish poor law commissioners, by the 1851 Medical Charities Act. This system has been characterised as 'the most comprehensive free medical care available in the British Isles'.[5] By the late nineteenth century, the first encounter that many Irishmen and women had with a qualified medical practitioner was through this dispensary system and scholars have rightly acknowledged its importance. Studies of the Irish dispensary system have almost exclusively focused on the official debate about the implementation and operation of the Medical Charities Act. While acknowledging the administrative and legislative limitations of the service, previous histories have portrayed the dispensary as the pathway along which the Irish were led to rational, scientific medicine and have suggested that rural communities enthusiastically accepted the service and the dispensary medical officers (MOs).[6] This article seeks to re-evaluate aspects of the service and suggests that the dispensary system was by no means as universally accepted as previous work has implied. As will be highlighted, the dispensary system encountered a series of problems, which hindered the delivery of healthcare at a local level. Issues of particular concern were access, pluralism and the need to discipline the dispensary MO. Patients' temporal and geographical access to a health service, and access conditions, is extremely influential in the delivery of healthcare. The problems associated with access, in part, ensured that heterodox medical practices remained the treatment of choice for some patients or were consulted alongside the dispensary system. However, as will be discussed, some patients did not turn to heterodox medical practices solely as a result of the limitations of the dispensary service; instead, they chose to do so as a result of their cultural preference for such medical therapeutics. Moreover, this preference was not confined to rural areas where MOs were less abundant and access was most limited. This article therefore, seeks to modulate and revise the history of the Irish dispensary service, without denying its fundamental importance. In doing so, it will explore some, though not all, of the issues that impacted upon the cultural acceptance of the service in the second half of the nineteenth-century.

The extant records of local medical dispensaries – minutes of dispensary district management committees, medical register books, vaccination books, and dispensary attendance and prescription books – have been consulted extensively in this study, as they facilitate an examination of the challenges that dispensary personnel encountered when delivering their services at a local level. The surviving dispensary records originate from approximately 10 per cent of the dispensary districts established under the 1851 Act. There is significant diversity in the type and quality of the material that has survived for each dispensary, while the rate of survival varied from district to district. The uneven pattern of survival of these records has resulted in their neglect. They are important, however, for any assessment of the local constraints under which the dispensary system operated, and of the issues that shaped the delivery of medical services within dispensary districts. Additional information on the dispensaries has been gleaned from board of guardian minutes published in local newspapers.

These sources have been supplemented by parliamentary reports, private correspondence, and published reminiscences of dispensary medical officers. The publications emanating from the Lady Dudley Nursing Scheme, which operated on the west coast of Ireland during the first decade of the twentieth century, have also been consulted. While all primary sources are biased in some respect, the annual reports of the Lady Dudley Nursing Scheme are particularly problematic. The Scheme was a philanthropic organisation, which, working with the Congested Districts Board, situated 'trained' nurses in the poorest parts of Ireland: the west and north-west.[7] Their published reports formed part of the organisation's propaganda drive to raise subscriptions and, when submitting their reports, the nurses, clearly conscious of their audience, may have exaggerated evidence. Despite these limitations, the reports constitute a rich source material, since the nurses assiduously included details of the difficulties they, and dispensary MOs, encountered when fulfilling their duties. Thus, rather than simply dismissing the reports outright, they have been used in conjunction with other primary material to explore the experiences of dispensary medical officers in the delivery of healthcare.

The Irish dispensary service

For many of Ireland's poor, the emergence of dispensary MOs facilitated greater access to regular medical provision. The origins of the service lay in eighteenth-century philanthropically funded dispensaries. The initial expansion of this service was encouraged in 1805 by legislation that allowed local grand juries and governors of county infirmaries to provide free medicine and medical assistance to the poor in dispensaries.[8] Dependent on local initiative and a fixed grant from grand juries presentments, which were tied to local subscriptions, the development of these dispensaries was piecemeal and their distribution irregular.[9] Unlike those in the county infirmaries, medical appointees to dispensaries were not obliged to possess any particular medical qualification and appointments were often 'bought'.[10] Although legislation passed in 1818 and 1836 improved the position of dispensaries and facilitated their expansion, their reputation, and that of the associated medical appointments, was mixed. In 1835, the *First Report from the Commissioners for Inquiring into the State of the Poorer Classes in Ireland* concluded that medical dispensaries inflated the numbers of patients in receipt of relief, were badly managed, and were sometimes established with the motivation of providing employment opportunities for medical attendants, rather than from an impulse to aid the poor. While the commissioners did admit to the existence of well-managed dispensaries attended by men of 'integrity and attention', the overall impression remained that the dispensary system was riddled with abuses.[11] The reform of the dispensary system soon became part of general investigations into the relief of the poor in Ireland in the 1830s and 1840s, a controversial and slow process.[12] In 1851 it was advanced with the passage of the Medical Charities Act, which brought the dispensaries within the remit of the poor law commissioners.[13]

Under the Act, Ireland was divided into 723 dispensary districts, comprising 960 separate dispensaries depots that would be served by 776 MOs and ten midwives.[14] Until the passage of the 1858 Medical Registration Act, candidates for dispensary posts were expected to hold a license from the Royal College of Surgeons in Ireland 'or the degree of some other "College or Body"'.[15] The poor law commissioners, however, retained the right to appoint individuals who did not meet these requirements.[16] As a consequence, prior to 1858, there was variation in the qualifications of dispensary MOs. The 1858 Medical Registration Act clarified the situation, as dispensary MOs were obliged to register under its terms.

The management of the new service was layered and bureaucratic. For the dispensary MO and the patients, the most immediate administrative body was the dispensary management committee, which was composed of local rate-payers resident in the district. The committee usually met fortnightly, although there is evidence that meetings became less fre quent over the course of the century.[17] Among other duties, the members were responsible for reviewing the dispensary records, ensuring the dispensary depot was maintained in good condition, and the selection of dispensary MOs, and in some cases, midwives. The poor law boards of guardians of the unions in which dispensaries were situated supervised the financial management of the dispensary and contracted with suppliers for approved medicines. The poor law guardians were also ex-officio members of the dispensary management committees. Decisions made by the guardians and the dispensary management committees, including the final appointment of medical staff, were subject to the approval of the poor law commissioners based in Dublin.

The duties of the dispensary MOs were onerous. They were expected to treat the range of ailments that presented among the patient population, while MOs' duties under the smallpox vaccination and nuisance removal legislation ensured that a large proportion of their time was devoted to tracking the occurrence of epidemic diseases.[18] Only patients with valid relief tickets qualified for treatment from dispensary MOs. These black (E-1) and red (E-2) relief tickets entitled patients to medical relief at the dispensary depot or in their home and were usually issued by dispensary committee members, or poor law relieving officers and wardens. Unlike other aspects of poor law welfare, patients' entitlement to the dispensary service was not precisely defined and, consequently, the individuals empowered to issue relief tickets determined applicants' entitlement to medical relief.[19] Ronald D. Cassell maintains that this resulted in a liberal interpretation of the phrase 'poor persons',[20] although individual dispensary committees attempted to monitor the issuing of relief tickets and in some cases cancelled them.[21] According to Cassell, the numbers of patients treated by the service expanded during the two decades after its establishment, rising from 11 per cent of the population in its first year of operation to 13 per cent in 1871–2. This occurred within the context of a declining population and there was some regional variation.[22]

In addition to the heavy workload, there were other constraints associated with the posts. After 1836 dispensary MOs were obliged to live within 'five miles of the institution' and this could prove problematic for the dispensary management committees.[23] Prospective candidates, who did not reside within the relevant district, were deemed ineligible for posts. In 1852 the incumbent MO at Delgany dispensary indicated his inability to 'comply with this condition' and the board of guardians was obliged to advertise for another MO.[24] The failure to reside in the district was cited in patients' complaints.[25] Dispensary MOs also experienced difficulties in locating suitable residences. Reminiscing in the 1870s of his experiences as a dispensary MO in Munster, Michael Joseph Malone recalled that his 'apartments were over a shop devoted to "public business", and as they had the advantage of being ceiled, a colony of rats took up their abode over them, under the shelter of the warm thatch'.[26] This difficulty in locating a residence of suitable quality was cited in demands for salary increases. In November 1853, Milltown dispensary management committee supported Samuel H. Agar's application for an increase on the basis that he had 'very great difficulty in finding a suitable residence'.[27] These constraints were also applied to the dispensary midwives and complaints were made when the midwife at Enniscorthy dispensary failed to comply.[28] While MO appointments were sought after, the heavy workload, and the hindrances and problems associated with filling positions ensured that vacant posts did not always attract a large number of candidates.[29] Practitioners with a viable private practice, such as Dr Dowes in Wexford in 1890, were in an enviable position in being able to decline dispensary work.[30]

Pluralism

Despite the heavy workload of dispensary MOs and other disadvantages associated with the posts, appointments to the newly established districts were highly prized. Ruth Barrington has suggested that by the later nineteenth century, as many as half of the registered medical practitioners in Ireland were employed as dispensary, workhouse or lunatic asylum medical officers.[31] An analysis of medical and trade directories indicates that in some counties the reliance on state-funded appointments was even greater. By 1892 in county Donegal approximately 65 per cent of the practitioners returned held state-funded appointments; the figure was nearer 75 per cent in county Mayo.[32] Dispensary posts provided MOs with a secure source of income and an important opportunity for new entrants into the medical profession, particularly in districts with limited scope for private practice. Consequently, the appointment of MOs frequently became contentious, particularly during the 1850s when the service was established and the initial tranche of medical posts was filled.[33] There were accusations that appointments were dependent on sectarian and political allegiances.[34] Having secured appointments, MOs wielded a significant amount of influence over the medical careers of colleagues. The incumbent dispensary MO was entitled to nominate individuals to act in *locum tenens* in their absence, thereby giving a colleague an entrée into the service.[35] These temporary appointments could become permanent when vacancies arose. At Enniscorthy dispensary, the incumbent MO, Thomas Cranfield, nominated Nicholas Furlong to act as his assistant in April 1871 when he was recovering from an illness.[36] The committee subsequently selected Furlong as MO following the death of Cranfield in February 1872. This system of nomination could be deployed to establish a monopoly on medical appointments within the control of one dispensary committee. The two MOs working at Enniscorthy dispensary until 1872, Cranfield and Samuel Goodison, established a system whereby they usually nominated each other when *locums* were required, strenuously resisting attempts to alter these arrangements.[37] In May 1863 Cranfield objected when the dispensary committee appointed Michael Sheridan to perform his duties during Cranfield's absence. Cranfield had omitted to nominate Goodison in his letter informing the committee he was incapacitated. Sheridan had recently qualified and was clearly anxious to secure a poor law appointment that would provide a degree of financial security.[38] Although Cranfield attempted to force the committee to overturn the appointment in Goodison's favour, the committee refused to oblige.[39] In the long term, the events did not undermine Cranfield's and Goodison's monopoly of Enniscorthy appointments, while Sheridan eventually secured a poor law position in the early 1880s, but as MO to Wexford workhouse. In addition to eliminating local competition, the arrangement had financial appeal. In February 1863, the dispensary management committee paid Goodison two guineas a week to perform Cranfield's duties during a fortnight's absence, adding to Goodison's existing salary as dispensary MO.

The attraction of appointments sometimes resulted in one individual holding several posts simultaneously. As a consequence, the poor law commissioners were obliged to regulate instances of pluralism and to consider its impact upon MOs' ability to deliver a system of healthcare. The commissioners were clearly concerned that MOs who held more than one post would become over-burdened, and that this, in turn, would adversely affect the quality of the service being delivered by the dispensary system. The commissioners actively discouraged dispensary MOs from engaging in pluralism, although the *General Rules for the Government of Dispensary Districts* did not specifically refer to the issue.[40] Thus, in July 1853, John George Battersby, MD, was obliged to resign his position at Milltown Dispensary in Mullingar poor law union, when he was appointed MO at Dungarvan workhouse in county Waterford.[41] Similarly, in 1852, a candidate for MO at the Delgany dispensary district in Rathdown poor law union, Richard McClelland, FRCSI, was deemed ineligible by the poor law commissioners, as he held a position in Newcastle dispensary district in Rathdrum Union.[42] The poor law commissioners, however, did not insist on a blanket ban on pluralism; rather, they assessed these cases on an individual basis. At the time of his appointment as MO at Wexford workhouse, Robert J. Crane, LRCSI, was already employed in a large dispensary district with a population of over 17,000. In this instance, the poor law commissioners 'did not think those duties could be properly performed' by one man and they refused Crane permission to hold both posts. In other cases, if the poor law commissioners felt the workload was not excessive, they allowed individuals to hold more than one post. Drs Thomas Rossiter and Peter Mullin were employed as both dispensary and workhouse MOs in county Wexford. The commissioners justified this on the basis that the population of the combined dispensary districts was only 19,778, while the workhouse duties 'devolved on two medical officers'.[43]

The efforts of the commissioners to reduce instances of pluralism were hampered by the limited number of registered practitioners operating in certain counties in Ireland. This was particularly true following the expansion in the number of medical posts established under the auspices of the poor law and other related welfare services later in the century. An analysis of seven counties – Dublin city and county; Cork city and county; Kilkenny city and county; Carlow town and county; Donegal; Kerry; and Mayo – in Ireland suggests that there was significant variation across Ireland in the number of registered doctors working in different counties.[44] These seven counties vary considerably in size and represent different social and economic contexts. Counties Donegal, Kerry and Mayo, in the west of Ireland, are the three largest counties and experienced periods of extreme economic hardship that, at times, verged on famine conditions during the nineteenth century.[45] Emigration from these counties was common and had become an accepted part of the 'life-cycle' amongst a declining population, as opportunities for employment were limited in a predominately rural economy.[46] The cities of Cork and Dublin had undergone some industrialisation in the nineteenth century and the emergence of tenement dwellings attests to the high levels of poverty in parts of these cities. The cities could also boast an expansion in middle class suburbs, such as Ranelagh and Rathmines in Dublin, and Blackrock in Cork.[47] A sizeable rural hinterland surrounded both cities. County Carlow and, particularly, county Kilkenny incorporated important provincial towns and, by the second half of the century, smaller industries such as brewing provided employment.[48] Outside the towns, the county economies remained largely dependent on agricultural activities.

The number of registered practitioners varied from county to county, but more tellingly the ratio of doctors to patients varied considerably from county to county and over time. Urban areas were generally better served; Dublin city could boast one registered practitioner to 478 people in 1861, while Cork had a ratio of one to 904. The provincial towns of Carlow and Kilkenny also fared reasonably well, although the number of regular practitioners declined in Carlow town between 1871 and 1891. None of the cities examined as part of this study have a ratio that is significantly worse than London, which, according to Anne Digby, approximated one doctor to 950 people in 1800 and, by 1841, the remainder of England 'was not far from this'.[49]

In Ireland, the number of registered practitioners declined after 1861. This occurred in the context of a decline in the size of the Irish population, falling from approximately 6.5 million in 1851 to just under 4.5 million in 1901, partly as a consequence of emigration. Medical practitioners formed part of this exodus.[50] As early as 1851, the number of physicians and surgeons in Ireland had declined by 14.4 per cent since 1841, although this figure includes the heavy toll exacted on medical practitioners by the Great Famine.[51] Greta Jones has suggested that this decline may, in part, have been caused by difficulties in maintaining a private practice in certain regions, ensuring that emigration proved attractive thereby contributing towards the depleted doctor/patient ratio.[52] There were significant regional differences in the rate of the decline. Dublin continued to be well-served by regular practitioners; its poorest doctor/patient ratio was recorded in 1901 when there was one registered practitioner to 679 people. None of the predominately rural districts among the seven counties even approximated the urban ratio. From 1861 to 1901 most counties outside Dublin and Cork struggled to maintain a ratio of one regular practitioner to c. 3,200 people. For example, in county Mayo in 1861 there were 14, 988 people per registered practitioner and this improved to 3,557 in 1901. This county had the poorest doctor/patient ratio. After 1871, the larger counties on the west coast, such as Kerry and Donegal, were not necessarily in a less favourable situation; in 1891, both counties Kerry and Donegal were better served than counties Carlow and Kilkenny. In counties Carlow and Kilkenny there was one regular practitioner to c. 3,824 and c. 3,629 people respectively, while in counties Kerry and Donegal there was one regular practitioner for a population of c. 3,380 and c. 3,201 respectively. Thus by the end of the nineteenth century, the doctor/patient ratio in Ireland had improved, in part, due to the decline in the size of the Irish population. However, registered practitioners and dispensary MOs in some parts of the country served an extremely large patient cohort throughout the nineteenth century and consequently their workload was onerous. In this context, the development of a uniform policy with regard to pluralism would have been difficult. The exigencies of the situation ensured that the poor law commissioners could merely monitor instances of pluralism among MOs, in a bid to protect the quality of the dispensary service. Pluralism became a relatively common practice, and remained so for the greater part of the nineteenth century.[53]

Distance and travel

The deliberations on pluralism and the 1836 residency stipulation highlight the importance that poor law commissioners, and other administrative bodies, assigned to securing patient access to the dispensary service. Such access varied depending upon the geographical proximity of patients to dispensary MOs and dispensary depots. Although the Irish transport infrastructure improved significantly during the nineteenth century,[54] the administrative bodies and their respective personnel were acutely conscious of the challenges that the size of dispensary districts presented. This was particularly pertinent as patients were only entitled to relief from the dispensary district in which they resided, rather than the depot nearest to their home. The size of the dispensary districts varied considerably across the country. For example, in New Ross poor law union, the estimated size of Dysartmoon dispensary district was over 21,416 statute acres (approximately 33.5 square miles) while The Rower district was only 7,062 statute acres (approximately 11 square miles). Undoubtedly this informed the decision to absorb the district of The Rower back into Dysartmoon in 1855. In Rathdown poor law union, Delgany district comprised 9,225 statute acres (approximately 14.5 square miles) and Powerscourt district was 21,601 statute acres (approximately 33.7 square miles). On the west and north-west coast, the size and variations were more striking. In the Glenties poor law union, Dungloe dispensary district comprised 72,220 statute acres (approximately 112.8 square miles), while Killybegs district was 12,625 statute acres (approximately 19.7 square miles).[55] Some of these districts were consolidated by 1872 as the number declined from 723 to 719.[56]

The size of dispensary depots ensured that Irish dispensary patients and MOs were often obliged to travel long distances, sometimes on foot, to access, in the case of patients, the dispensary committee members for relief tickets, the dispensary MOs and the dispensary depot itself, and, in the case of MOs, to access patients' homes and the depots in his district.[57] The bodies that administered dispensaries were acutely conscious of these problems and spent a large proportion of their time in attempting to ameliorate the difficulties. The location of dispensaries, their opening times, and the days and times of the MOs' attendance were subjects of endless debate at dispensary committee meetings and communications with the poor law commissioners, as the commissioners sought to impose some uniformity. Generally, an MO attended each dispensary depot twice in one week. In Delgany dispensary district it was agreed there would be two depots – one in the village of Delgany and the second in Kilmurray village. In March 1852, the MO, Henry Pelim Browne, MD and LRCSI, was required to attend at each dispensary three times a week from 11am. Within a month the committee suggested that his attendance at each dispensary be reduced to two days a week – from 11am in Delgany and 1pm in Kilmurray.[58] In 1873, the Enniscorthy management committee required the MO to attend Oylgate and Clondaw dispensary depots at 12 noon, an hour later than the depot at Enniscorthy, due to the long distance applicants were obliged to travel.[59] The management committee of Kilmore dispensary, in Monaghan poor law union, originally established two depots in their district, at Smithboro and Slieveroe, but in 1886 the number was increased to three depots. The addition of an extra depot coincided with a decision to reduce the number of days the MO was required to attend Slieveroe to one day a week.[60] The decision of Kilmore management committee reflected an attempt to reach a balance between Browne's workload and patients' access to the depot, which the committee felt was better served by improving geographical rather than temporal access. For patients, this effectively meant the local dispensary was accessible for approximately two hours every three or four days. As required under the Medical Charities Act, once agreed upon, the places and times of dispensary opening hours, in addition to the names of the MOs, were advertised at the dispensary to inform the public.[61] The awareness of the inherent difficulties in ensuring patients' access was further reflected in the frequent discussions of the location of the meetings of dispensary management committees. For example, in June 1853, when the poor law commissioners expressed their disapproval of Delgany dispensary management committee's habit of holding meetings in locations other than Delgany village, the committee members insisted that their preferred location, Glen Cottage, facilitated patient access. Unlike the village of Delgany, Glen Cottage was halfway between the two dispensary depots. It also had the added bonus of being rent-free. The committee stubbornly resisted all attempts to change the location until August 1857, when they finally agreed to hold their meetings in Delgany village.[62]

Similarly, patients' ability to gain physical access to dispensary MOs and to the members of the dispensary committee was important. When attempting to secure a home visit, for relatives too ill to appear at the dispensary in person, families or friends of the ill were obliged to obtain a red relief ticket (E-2). These tickets were subsequently presented to the dispensary MO, who was expected to visit the patient. In a bid to facilitate families' access to the various individuals empowered to issue dispensary tickets, the poor law commissioners issued a circular in March 1853 entreating local poor law boards of guardians to ensure that such individuals resided in 'different localities in several parts of the district so that sufficient facility for obtaining such relief with due convenience to the sick poor resident in all parts of the district may be afforded'.[63] The timing of the next stage in the procedure – the presentation of visiting tickets to dispensary MOs – was also a matter of concern. In 1857, the poor law commissioners requested that management committees select a time for the presentation of these tickets at the home of the MO, in order to allow dispensary MOs to visit the home of the patient on the same day. The commissioners were opposed, however, to the decision made by some management committees that necessitated the presentation of relief tickets at the home of the doctor before 9am, as in 'most rural dispensary districts' the selection of 'such an early hour' would 'render it impracticable for the relatives or friends of the sick poor residing at a distance from the residence of the medical officer to obtain and present a visiting ticket'.[64] These efforts on the part of the poor law commissioners and local dispensary management committees to overcome the difficulties caused by the geographical distance between dispensary patients and MOs, the personnel empowered to issue relief tickets, and the dispensary depots themselves, occupied a central place in debates about patient usage of the dispensary service. Each administrative body recognised the importance of overcoming the difficulties presented by distance although, as indicated above, they responded in different ways.

The problems that distance and securing suitable transport presented were also a cause of concern for the dispensary MOs themselves. As Irvine Loudon and Anne Digby have shown, by the middle of the eighteenth century, doctors' capacity to travel and the geographical distribution of patients was an important consideration in the maintenance of a successful private practice.[65] These factors were equally important to dispensary MOs. The challenges that the terrain of rural Ireland presented to them frequently featured in the dispensary management committees' campaigns to secure salary increases for dispensary MOs. In 1854, the Delgany management committee supported Dr Browne's request for a salary increase – from £75 to £100 – on the basis of the 'great extent and mountainous character of the district', which forced 'the medical officer to keep a horse'. They also cited 'the additional duty imposed by the 2 dispensary depots and the very small additional income in respect to private practice'.[66] As Loudon has shown, the possession of a reliable and healthy horse was important for a rural practitioner, but 'visiting by horseback was a hard and exhausting life'.[67] It could also be dangerous; George Lang, MO at Killann dispensary in the Enniscorthy poor law union met with a serious accident following a fall from his horse while on duty in July 1861.[68] In his reminiscences of a brief period spent as a dispensary MO in Munster – the date is not specified – M.J. Malone expressed some bitterness at the distances and conditions of travel that the rural dispensary MO was obliged to endure when compared to his urban counterpart, claiming that 'the doctor ... has often to leave his bed to answer the sick-call on foot, over mountain and bog, depending on his way on the instincts rather than the vision of the guide who faithfully steers him through the rayless darkness of his destination'.[69] In this type of rural terrain, it is unlikely that coaches were used with great frequency outside of towns.

The problems associated with transport and access continued to cause concern at the end of the nineteenth century, despite the improvements made to transport infrastructure.[70] It remained a particular problem in the west of Ireland in the first decade of the twentieth century. Then, the annual reports of the Lady Dudley Nursing Scheme included considerable commentary on the travails dispensary MOs encountered while on duty. Although they were not formally connected with the dispensary system, the annual reports of that organisation contain nurses' accounts of their experiences working in dispensary districts. The distances travelled and obstacles encountered in the nurses' own journeys to attend the homes of patients are recorded. To aid transport, the nurses supplemented the use of horses with bicycles, reflecting a transitory trend in medical transport identified by Loudon and the increase in popularity of bicycles for women.[71] They were also obliged to travel by boat to visit patients on the various islands scattered along the western coast.[72] The nurses frequently reported on the difficulties that dispensary MOs encountered when travelling to the homes of their patients. In one case the nurse ordered the family of a patient to seek an MO who was in attendance at his dispensary situated 14 miles from the patient's home. She recorded that it was about '12 o'clock noon when he had been telegraphed for, but was not able to reach the patient before 6pm'.[73] In another case the nurse sent for a doctor who lived nine miles away to attend a woman in labour. The patient's 'husband had first to walk a mile to get a horse; on the way the horse cast a shoe. The man had to return some distance to the forge and being Sunday there was of course delay about the fire. The doctor did not arrive on time, but the patient recovered.'[74] While the Dudley nurses were evidently anxious to emphasise the need for their own services and, therefore, to underline the limitations of the dispensary service, there is little doubt that doctors continued to encounter significant problems when required to 'promptly' attend patients.

Engagement

It would, however, be a mistake to assume that the elimination of the multiple problems associated with access to the Irish dispensary service would have guaranteed patients' engagement with the service or their acceptance of treatments. Although Cassell asserts that the dispensaries introduced medical science and successfully challenged traditional ideas, attitudes, techniques, and procedures in rural Ireland,[75] there is abundant evidence to suggest that patients' responses to the dispensary system and its associated services were complex. In districts where 'formal' medical practitioners were both available and accessible, patients continued to consult heterodox practitioners as an alternative to bio-medicine, or in conjunction with it. In these instances, patients expressed their agency through a hesitancy to avail of a free service.

The continued popularity of heterodox practitioners in nineteenth and twentieth-century Ireland is well documented.[76] In 1910, the *Report as to the Practice of Medicine and Surgery by Unqualified Persons in the United Kingdom* remarked upon the presence of heterodox practitioners in almost all dispensary districts and highlighted the diversity of attitudes among dispensary MOs towards particular groups of non-registered practitioners.[77] While the authors of the explanatory report that accompanied the evidence were adamant that the practices of chemists, bonesetters, handywomen, and others were invariably harmful, the replies of the dispensary MOs from 717 districts, which formed the basis of the report, are less homogeneous. Among dispensary MOs, there was general agreement that chemists, who performed minor operations, dentistry, and supplied drugs, were dangerous, and represented a real loss of income to registered practitioners. Loudon identified a similar hostility towards druggists in the early nineteenth century.[78] The activities of bonesetters, handywomen, and herbalists were also criticised and bonesetters, in particular, were accused of permanently disabling patients. A significant number of dispensary MOs, however, described traditional practitioners in less hostile terms, while dispensary MOs in Belmullet, county Mayo, maintained that in some instances 'the result was decidedly good'. For these doctors the harm lay not with the consultation of 'folk' practices *per se* but with the failure to consult a dispensary MO when a case became serious.

As I have suggested elsewhere, dispensary MOs' attitudes towards 'irregular' practitioners' could be quite nuanced.[79] Evidence of patients' preference for 'irregular' practitioners in cases where 'formal' medical assistance was available can be found in numerous sources. M.J. Malone's reminiscences included a detailed description of the activities and popularity of renowned local healer Biddy Early, [80] while the correspondence and publications of nineteenth-century antiquarians are redolent with details of such practitioners and their therapeutics.[81] Much to the chagrin of the poor law commissioners, smallpox inoculators continued to be consulted despite the availability of smallpox vaccination, free of charge, through the dispensary service.[82] The Lady Dudley nurses found the adherence to these practitioners particularly frustrating. One Dudley nurse became very irate when she discovered that her treatment for an ulcerated leg had been discarded in favour of a moss cure prescribed by a handywoman.[83] The nurses encountered similar resistance when they attempted to cope with other emergencies such as broken bones, burns and childbirth.[84] The patients had either rejected the free services of the nurse or they were willing to apply several treatments simultaneously, a response not unique to Ireland.[85]

Dispensary MOs encountered other challenges in establishing themselves and the service among the poor population in rural districts. Too often, the dispensary depots were unappealing or uncomfortable places for patients. Dispensary management committees failed to maintain the buildings and these were sometimes allowed to become quite dilapidated.[86] The waiting-room at Enniscorthy dispensary was without a fire until 1901, while its proximity to the fever hospital dissuaded patients from attending during an outbreak of smallpox in 1895.[87] Dispensary MOs were obliged to negotiate their position of influence with the local clergy. Potential candidates for state-funded appointments attempted to establish moral, as well as medical, credentials in their applications for positions, and often provided testimonials from members of the religious, as well as the medical, establishment when seeking a position. When the post of manager to Carlow Lunatic Asylum became vacant in August 1834, Paul Cullen, accoucheur and apothecary in Carlow town, provided testimonials from the county's Roman Catholic clergy and from the senior academics of Carlow College.[88] Similarly, in August 1855, Michael Walsh, the parish priest for New Ross, supported the candidature of Abraham Alcock as the MO for Dysartmoon dispensary.[89]

In addition to securing testimonials from members of the clergy when applying for a post, at times the dispensary MOs enlisted the influence of the priest to persuade a patient to adhere to their recommendations or to attend the local hospital. An MO in county Donegal, suspecting a case of typhus, attempted to have a small boy 'removed' to hospital, but when the family resisted, the local parish priest was called upon to 'interfere'.[90] The local clergy sometimes displayed some ambivalence towards dispensary MOs. In their letters requesting the support of nurses from the Lady Dudley organisation, Catholic parish priests emphasised both the limitations of the location of dispensary MOs and the problems associated with modes of transport and communication. They were sometimes quite critical of the MOs. This tension between the priest and dispensary MO may have been partly the result of confessional differences. By 1871, 'just over three-quarters of the Irish population were Catholics; 12 per cent were members of the Church of Ireland and a further 9 per cent were Presbyterians', while it is estimated that only 34 per cent of doctors were Catholic in 1870.[91] Although the proportion of Catholic appointments increased, the priest remained an influential 'confidant and advisor in the sick room'[92] and his role was not confined to matters pertaining to reproduction.

Conclusion

Some patients continued to use the services of non-registered practitioners, in preference to the free services of the local dispensary MO, in part because of the limitations of the service, particularly the problems associated with the size of the districts, the distribution of services, and the mechanisms of securing relief tickets. Ensuring access to the dispensary service was a matter of serious concern for the different administrative bodies and this concern shaped the delivery of the service at local level, particularly in relation to decisions as to the location of dispensary depots and MOs' hours of attendance. The improvements made to the road networks and transport innovations in the late nineteenth century only partly alleviated these problems. The expansion of the railway and the popularity of bicycles reduced, but did not replace, the practitioners' reliance on horseback as a reliable means of transport and, for the 'poor persons' referenced in the 1851 Act, their limited economic resources would have prohibited expenditure on transport.[93]

Hesitancy in using the services, however, was not determined by these problems alone. The dispensary MO and, by extension, bio-medicine itself, continued to occupy a problematic position in Ireland by the first decade of the twentieth century. Patients had by no means 'developed an exclusive preference for a particular therapy'[94] and the free services of dispensary MOs were not always the first port of call. In the pluralistic medical world of the nineteenth and early twentieth centuries, dispensary MOs' authority in matters of health had not been absolutely established. They were obliged to negotiate their position and authority within rural communities with other well-established figures, including the local priest and the practitioners delineated as 'folk' and 'traditional'. The dispensary MOs' responses to these negotiations were diverse. While the 'folk' practitioners were criticised by some, others tolerated 'traditional' therapeutics that occupied a secure place within a community's medical repertoire.[95] In addition, the local priest successfully maintained his position of influence and in the sick-room he became both an ally and a rival to the dispensary MO. By the beginning of the twentieth century, the dispensary service occupied an important but contested place within Irish communities.

Notes

* I would like to acknowledge the generous support of the Wellcome Trust and Professor Greta Jones in the preparation of this article.

1 Ian Mortimer, 'The rural medical marketplace in Southern England c. 1570–1720', in Mark S.R. Jenner and Patrick Wallis (eds), *Medicine and the Market in England and Its Colonies, c. 1450–c. 1850* (Basingstoke, 2007), p. 69.
2 Oliver MacDonagh, 'Ideas and institutions, 1830–45', in W.E. Vaughan (ed.), *A New History of Ireland V: Ireland Under the Union, 1801–1870, I* (Oxford, 1989), p. 210; idem., *Ireland* (New York, 1968), p. 27; Gearoid Ó Tuathaigh, *Ireland Before the Famine 1798–1848* (Dublin, 1972), p. 95; Cormac Ó Gráda, *Ireland: A New Economic History 1780–1939* (Oxford, 1994), p. 97.
3 Timothy P. O'Neill, 'Fever and public health in pre-Famine Ireland', *Journal of the Royal Society of Antiquaries of Ireland*, 103 (1973), pp. 1–34.
4 For a discussion of the use of the term see Waltraud Ernst, 'Plural medicine, tradition and modernity. Historical and contemporary perspectives: View from below and from above', in idem. (ed.), *Plural Medicine, Tradition and Modernity, 1800–2000* (London, 2002), pp. 1–19; Roger Cooter (ed.), *Studies in the History of Alternative Medicine* (London, 1988); Matthew Ramsey, 'Magical healing, witchcraft and elite discourse in eighteenth and nineteenth-century France', in Marijke Gijswijt-Hofstra, Hilary Marland and Hans de Waardt (eds), *Illness and Healing Alternatives in Western Europe* (London, 1997), pp. 14–37.

5   Ronald D. Cassell, *Medical Charities, Medical Politics. The Irish Dispensary System and the Poor Law, 1836–1872* (Woodbridge, 1997), p. 128. The 1851 Medical Charities Act did not include the county infirmaries and district fever hospitals. These would be subsequently incorporated. See Cassell, *Medical Charities, Medical Politics*, pp. 103–8.

6   Ruth Barrington, *Health, Medicine and Politics in Ireland 1900–1970* (Dublin, 1987), pp. 8–12; Helen Burke, *The People and the Poor Law in Nineteenth-Century Ireland* (Littlehampton, 1987), pp. 243–82; Cassell, *Medical Charities, Medical Politics*, pp. 78–108.

7   Ann Wickham, '"She must be content to be their servant as well as their teacher": The early years of District Nursing in Ireland', in Gerard M. Fealy (ed.), *Care to Remember: Nursing and Midwifery in Ireland* (Cork, 2005), pp. 102–21; Ciara Breathnach, *The Congested Districts Board of Ireland, 1891–1923: Poverty and Development in the West of Ireland* (Dublin, 2006).

8   County Infirmaries (Ireland) Act (45 Geo. iii c.111 (1805)).

9   Cassell, *Medical Charities, Medical Politics*, p. 8; MacDonagh 'Ideas and institutions', p. 208.

10 Laurence M. Geary, *Medicine and Charity in Ireland 1718–1851* (Dublin, 2004), p. 131.

11 *First Report from the Commissioners for Inquiring into the State of the Poorer Classes in Ireland*, Part ii, Appendix b, 1835 [369] xxxii, p. 3.

12 For a detailed discussion of the issues involved see Cassell, *Medical Charities, Medical Politics*, pp. 18–77.

13 Ibid., p. 78.

14 Ibid., p. 92.

15 Ibid., p. 90.

16 An Act to provide for the better Distribution, Support, and Management of Medical Charities in Ireland; and to amend an Act of the Eleventh Year of Her Majesty, to provide for the Execution of the Laws for the Relief of the Poor in Ireland (14 & 15 Vic. c. 68 (1851)).

17 For example see Enniscorthy Dispensary Minute Book, 28 November 1854–13 April 1897, WX/BG 87/1/3/1, Wexford County Archives Service (hereafter WCA).

18 See the *Annual Reports of the Commissioners for Administrating the Laws for Relief of the Poor in Ireland under the Medical Charities Act*.

19 Cassell, *Medical Charities, Medical Politics*, p. 95.

20 Ibid.

21 For examples, see Enniscorthy Dispensary Minute Book, 9 August 1859 and 22 April 1864, WX/BG 87/1/3/1, WCA.

22 Cassell, *Medical Charities, Medical Politics*, p. 96.

23 Geary, *Medicine and Charity*, p. 136; Enniscorthy Dispensary Minute Book, 13 October 1863, WX/BG 87/1/3/1, WCA.

24 Delgany Dispensary Minute Book, Ms 16411, Hodson papers, National Library of Ireland (hereafter NLI).

25 Chief Secretary Office Registered Papers (hereafter CSORP), 1859/11304, National Archives of Ireland (hereafter NAI).

26 M.J. Malone, 'Recollections of a country dispensary', *Irish Monthly*, 6 (February 1878), p. 73.

27 Miltown Dispensary Minutes, Pos 9551, NLI.

28 Enniscorthy Dispensary Minute Book, 10 June 1896, WX/BG 87/1/3/1, WCA.
29 Miltown Dispensary Minutes, Pos 9551, NLI.
30 Enniscorthy Dispensary Minute Book, 25 February 1890, WX/BG 87/1/3/1, WCA.
31 Barrington, *Health, Medicine and Politics*, p. 16.
32 Catherine Cox, 'The medical marketplace and medical tradition in nineteenth-century Ireland', in Stuart McClean and Ronnie Moore (eds), *Folk Healing and Health Care Practices in Britain and Ireland: Stethoscopes, Wands and Crystals* (Oxford, 2010), pp. 55–79; *Medical Directory for Ireland, 1852–1902.*
33 For example see CSORP 1856/11789, NAI.
34 'Carlow union election of medical officer', *Freeman's Journal*, 23 January 1877; 'Protestant ascendancy in Carlow', *Freeman's Journal*, 30 January 1877.
35 *Medical Directory for Ireland, 1864.*
36 Enniscorthy Dispensary Minute Book, 27 April 1871 and 20 February 1872, WX/BG 87/1/3/1, WCA.
37 For some examples see Enniscorthy Dispensary Minute Book, 23 February 1863, 18 November 1865, 24 February 1869, WCA, WX/BG 87/1/3/1.
38 *Medical Directory for Ireland, 1889*, p. 165.
39 Enniscorthy Dispensary Minute Book, 3 June 1863, WX/BG 87/1/3/1, WCA.
40 *General Rules for the Government of Dispensary Districts*, Kilmore Dispensary papers, P42/77, UCD Archives (hereafter UCDA).
41 Miltown Dispensary Minutes, Pos 9551, NLI.
42 Delgany Dispensary Minute Book, Ms 16411, Hodson papers, NLI. See also CSORP 1856/11789, NAI and *Medical Directory for Ireland, 1852.*
43 Letter from Stanley, Secretary, Poor Law Commissioners to Edward Horsman, 26 January 1856, CSORP 1856/11789, NAI.
44 The sources used to compile these figures include the *Medical Directory for Ireland* (before and after the 1858 Medical Act), nineteenth-century city and county trade directories, and university rolls containing details of student entrants. These have been supplemented by numerous parliamentary reports, including census returns, and the annual reports of poor law commissioners and of the lunacy inspectorate. These include 1841, 1851, 1861, 1871, 1881, 1891 and 1901 census returns, published returns and accounts of the poor law commissioners, the lunacy inspectorates, the Apothecary Hall, and the committees established to inquire into the condition of the poor in Ireland and the medical charities. The returns for apothecaries have been gleaned from trade directories. Although the majority claimed to be licentiates of the Apothecary Hall, a small number of self-styled practitioners were undoubtedly active. There are significant limitations with the source material, particularly when using the *Medical Directory for Ireland*. The returns were dependent on individual medical practitioners submitting completed entries themselves; this did not always occur and compilers of the directories were obliged to use out-dated details.
45 Timothy P. O'Neill, 'The persistence of famine in Ireland', in C. Póirtéir (ed.), *The Great Irish Famine* (Cork, 1995), pp. 204–18.
46 For a discussion of the impact of emigration on nineteenth-century Ireland, see David Fitzpatrick, *Irish Emigration 1801–1921* (Dublin, 1990); idem.,

'Emigration, 1801–70', in Vaughan (ed.), *New History of Ireland V*, pp. 585–8; Timothy Guinnane, *The Vanishing Irish: Households, Migration and the Rural Economy in Ireland 1850–1914* (Princeton, 1997).

47 For useful histories of Dublin urban development, see Mary E. Daly, *Dublin: The Deposed Capital. A Social and Economic History 1860–1914* (Cork, 1984); Jacinta Prunty, *Dublin Slums, 1800–1925: A Study in Urban Geography* (Dublin, 1999); F.H.A. Aalen and Kevin Whelan (eds), *Dublin City and County from Prehistory to Present: Studies in Honour of J. H. Andrews* (Dublin, 1992). For Cork see A.M. Fahy, 'Place and class in Cork', in Patrick O'Flanagan and Cornelius G. Buttimer (eds), *Cork: History and Society. Interdisciplinary Essays on the History of an Irish County* (Dublin, 1993), pp. 793–812 and David Dickson, *Old World Colony: Cork and South Munster, 1630–1830* (Cork, 2005).

48 For histories of Carlow and Kilkenny, see William Nolan and Kevin Whelan (eds), *Kilkenny: History and Society. Interdisciplinary Essays on the History of an Irish County* (Dublin, 1990) and Thomas McGrath (ed.), *Carlow: History and Society. Interdisciplinary Essays on the History of an Irish County* (Dublin, 2008).

49 Anne Digby, *Making a Medical Living. Doctors and Patients in the English Market for Medicine, 1720–1911* (Cambridge, 2002), p. 19.

50 Greta Jones, '"Strike out boldly for the prizes that are available to you": Medical emigration from Ireland 1860–1905', *Medical History*, 54:1 (2010), pp. 55–74.

51 W.E. Vaughan, 'Ireland c. 1870', in Vaughan (ed.), *New History of Ireland V*, pp. 738–9.

52 Jones, 'Strike out boldly for the prizes that are available to you'.

53 For example see *Medical Directory for Ireland, 1892*.

54 Cormac Ó Gráda, 'Industry and communications, 1801–45', in Vaughan (ed.), *New History of Ireland V*, pp. 137–55 and H.G. Gribbon, 'Economic and social history, 1850–1921', in W.E. Vaughan (ed.), *New History of Ireland VI: Ireland under the Union, II* (Oxford, 1996), pp. 260–356.

55 *First Annual Report of the Commissioners for Administrating the Laws for Relief of the Poor in Ireland under the Medical Charities Act*, 1852–53, [1609] 1, Appendix b, table 1, pp. 88–133.

56 Cassell, *Medical Charities, Medical Politics*, p. 93.

57 Ian Mortimer has highlighted the dangers of thinking in terms of a static patient cohort in his analysis of a rural medical marketplace. See Mortimer, 'The rural medical marketplace in southern England', pp. 70–4.

58 Delgany Dispensary Minute Book, Ms 16411, Hodson papers, NLI.

59 Enniscorthy Dispensary Minute Book, 11 March 1873, WX/BG 87/1/ 3/1, WCA.

60 Kilmore Dispensary papers, P42/42–43, UCDA.

61 An Act to provide for the better Distribution, Support, and Management of Medical Charities in Ireland; and to amend an Act of the Eleventh Year of Her Majesty, to provide for the Execution of the Laws for the Relief of the Poor in Ireland (14 & 15 Vict. c.68 (1851)).

62 Delgany Dispensary Minute Book, Ms 16411, Hodson papers, NLI and Miltown Dispensary Minutes, Pos 9551, NLI.

63 Circular to Boards of Guardians, 26 March 1853, *Second Annual Report of the Commissioners for Administering the Laws for Relief of the Poor in Ireland, under the Medical Charities Act*, 1854 [1759] xx, Appendix a, no. 2, pp. 24–5.

64 Kilmore Dispensary papers, P42/108, UCDA.
65 Digby, *Making a Medical Living*, pp. 111–17; idem., *The Evolution of British General Practice 1850–1948* (Oxford, 1999), pp. 145–9; Irvine Loudon, 'Doctors and their transport, 1750–1914', *Medical History*, 45:2 (2001), pp. 185–206.
66 Delgany Dispensary Minute Book, Ms 16411, Hodson papers, NLI; Kilmore Dispensary papers, P42/37–38, UCDA.
67 Loudon, 'Doctors and their transport', p. 205.
68 'Enniscorthy poor law union', *Enniscorthy News*, 6 July 1861.
69 M.J. Malone, 'Doctoring under difficulties', *Irish Monthly*, 7 (July 1879), pp. 383–4.
70 According to Cormac Ó Gráda, 'between 1801 and 1845 grand juries spent £300,000 to £400,000 annually on road repair and construction', see Ó Gráda, 'Industry and communications, 1801–45', p. 146.
71 Loudon, 'Doctors and their transport', p. 199; Brian Griffin, *Cycling in Victorian Ireland* (Dublin, 2006), pp. 102–41; *Second Annual Report of Lady Dudley's Scheme for the Establishment of District Nurses in the Poorest Parts of Ireland April 1904–April 1905*, p. 6.
72 *Second Annual Report of Lady Dudley's Scheme for the Establishment of District Nurses in the Poorest Parts of Ireland 1904–1905*, p. 9.
73 *Ninth Annual Report of Lady Dudley's Scheme for the Establishment of District Nurses in the Poorest Parts of Ireland 1911*, p. 18.
74 *Fifth Annual Report of Lady Dudley's Scheme for the Establishment of District Nurses in the Poorest Parts of Ireland 1907*, p. 10.
75 Cassell, *Medical Charities, Medical Politics*, p. 10.
76 See Patrick Logan, *Irish Country Cures* (Belfast, 1981) and Cox, 'The medical marketplace and medical tradition'.
77 *Report as to the Practice of Medicine and Surgery by Unqualified Persons in the United Kingdom*, 1910 [Cd.5422] xliii, pp. 65–86.
78 Irvine Loudon, 'The vile race of quacks with which this country is infected', in W.F. Bynum and R. Porter (eds), *Medical Fringe and Medical Orthodoxy* (London, 1987), pp. 106–28.
79 Cox, 'The medical marketplace and medical tradition'; J. Bradley, 'Medicine on the margins? Hydropathy and orthodoxy in Britain, 1840–60', in Ernst (ed.), *Plural Medicine, Tradition and Modernity*, pp. 19–39.
80 Malone, 'Recollections of a country doctor', p. 77. For more information on Biddy Early see Gearóid Ó Crualaoich, *The Book of the Cailleach: Stories of the Wise-Woman Healer* (Cork, 2003).
81 See for example papers of Eugene O'Curry (1796–1862), UCDA; Thomas More Madden, 'Revival of old fallacies bearing on medicine', *Dublin Journal of Medical Science*, 90 (1890), pp. 22–41; W.R. Wilde, 'Irish popular superstitions', *Dublin University Magazine* (May 1849–May 1850); idem., *The History of Irish Medicine and Popular Cures* (Dublin, n.d.); Henry S. Purdon, 'Notes on old native remedies', *Dublin Journal of Medical Science*, 100 (1895), pp. 214–18; idem., 'Old Irish "herbal" skin remedies', *Dublin Journal of Medical Science*, 106 (1898), pp. 27–31.
82 *Second Annual Report of the Commissioners for Administering the Laws for Relief of the Poor in Ireland, under the Medical Charities Act*, 1854 [1759] xx, appendix b, no. 5, p. 29; *Third Annual Report of the Commissioners for Administering the Laws for Relief of the Poor in Ireland, under the Medical Charities Act*,

1854–55 [1908], xvi, pp. 12, 78–9; *Fourth Annual Report of the Commissioners for Administering the Laws for Relief of the Poor in Ireland, under the Medical Charities Act*, 1856 [2062] xix, pp. 78–9; Deborah Brunton, 'The problems of implementation: The failure and success of public vaccination against smallpox in Ireland, 1840–1873', in Elizabeth Malcolm and Greta Jones (eds), *Medicine, Disease and the State in Ireland, 1650–1940* (Cork, 1999), pp. 138–57.

83   *Third Annual Report of Lady Dudley's Scheme for the Establishment of District Nurses in the Poorest Parts of Ireland April 1905–April 1906*, p. 13. For a discussion of nurses' attitudes towards 'traditional' African medicine and their role as 'cultural brokers' see Ann Digby and Helen Sweet, 'Nurses as cultural brokers in twentieth-century South Africa', in Ernst (ed.), *Plural Medicine, Tradition and Modernity*, pp. 113–29.

84   *Third Annual Report of Lady Dudley's Scheme for the Establishment of District Nurses in the Poorest Parts of Ireland April 1905–April 1906*, p. 14.

85   For a detailed discussion of the scholarship on this topic see Marijke Gijswijt-Hofstra, Hilary Marland and Hans de Waardt, 'Introduction. Demons, diagnosis and disenchantment', in idem. (eds), *Illness and Healing Alternatives in Western Europe*; Ernst (ed.), *Plural Medicine, Tradition and Modernity*; Roberta Bivins, *Alternative Medicine? A History* (Oxford, 2007).

86   For example see Enniscorthy Dispensary Minute Book, 13 April 1895, WX/BG 87/1/3/1, WCA.

87   Enniscorthy Dispensary Minute Book, 25 January 1895 and 24 October 1901, WX/BG 87/1/3/1, WCA.

88   See also Chief Secretary's Office, Official Papers, 1835/346 (hereafter OP), NAI, see also OP 1835/343, 344, 347, 348, NAI.

89   Letter from Michael Walsh PP, New Ross, to the Poor Law Commissioners, 18 August 1855, CSORP 1856/11789, NAI.

90   *Eight Annual Report of Lady Dudley's Scheme for the Establishment of District Nurses in the Poorest Parts of Ireland 1910*, p. 16; *Ninth Annual Report of Lady Dudley's Scheme for the Establishment of District Nurses in the Poorest Parts of Ireland c.1912*, p. 19.

91   Vaughan, 'Ireland c.1870', pp. 738–9.

92   Digby, *Evolution*, p. 233.

93   Loudon, 'Doctors and their transport'.

94   Gijswijt-Hofstra, Marland and Waardt, 'Introduction. Demons, diagnosis and disenchantment', p. 5.

95   Ibid., pp. 7–8.

4
Suicide and Insanity in Post-Famine Ireland

Georgina Laragy

Introduction

The relationship between suicide and insanity in nineteenth-century Ireland can be traced back to two specific developments: firstly, the legal link between suicide and madness that had been established in the medieval period; and secondly, the interpretation of insanity as a mental illness subject to medical intervention.[1] These issues are crucial for our understanding of attitudes to suicide in post-Famine Ireland. The relationship between suicide and insanity was first codified within the legal canons of medieval England and, by the nineteenth century, the role of the medical profession in caring for those suffering from insanity had been established. By this time, however, the legal position of suicide was beginning to change. This article asks whether, in the face of these changes, the verdict of temporary insanity, attributed to a suicide, continued to have any relevance, and to what extent did doctors, government officials and ordinary men and women continue to link suicide with insanity. Firstly, the legal background, and changes, in the nineteenth century will be explored, along with the attitudes of the churches that punished suicides posthumously. This section also reveals a possible connection between concealment of suicides and sanity. Secondly, the article will explore how decisions were made by the family and friends of those who were suicidal. Did they call doctors when someone threatened suicide? Given that attempted suicide was also a crime, did families call the police when someone attempted suicide?

Was the verdict of temporary insanity an historical remnant from a time when communities and families hoped to evade the punishment, legal and ecclesiastical, that suicide warranted? Or had the link between suicide and insanity pervaded popular understandings of suicidal behaviour, to the extent that it influenced decisions made by Irish Victorians and Edwardians faced with suicidal family members and friends? In order to examine the question of whether or not the relationship between suicide and insanity existed beyond the legal question of innocence or guilt, various legal sources are used, including coroners' inquests and prisoner petitions. Inquests provide testimony from people who were dealing with the aftermath of a successful suicide. They describe behaviour and action taken before the deceased had committed the lethal act. Prisoner petitions contain pleas from those who were convicted of attempted suicide and imprisoned. These documents contain decisions made by judges on custodial sentences and provide an interesting perspective on their rationalisation of such sentences. Material found in inquests and petitions is supported by information contained within government reports, newspapers, medical treatises and other writings.

Suicide and insanity: A legal loophole?

According to the legal code, suicide was a crime in Ireland until 1993. Historically known as self-murder or *felo de se*, it was punishable by ignominious burial until 1823 and the forfeiture of goods and chattels until 1872.[2] Despite these changes suicide remained a crime without legal punishment. In order to be judged innocent of self-murder the deceased had to be found temporarily insane at the time of the 'rash act'. This exemption for 'madmen' had been in place within the English legal code since 1000 A.D. Following the Anglo-Norman invasion of Ireland in the twelfth century, the English view of suicide was replicated in Ireland in parts of the country where their authority was imposed. In the medieval period this crime was investigated by the coroner who held an inquest. The duties of the coroner, which included the investigation of all sudden deaths, were set out in detail in the statute *De Officio Coronatoris* 1276 ('Of what things a coroner shall inquire') and were applied to Ireland under the provisions of Poynings Law.[3] An eighteenth-century manual for local government officials in Ireland stated that coroners

> shall go to the Places where any be slain, or suddenly dead or wounded ... shall inquire in this matter ... in like manner it is to be inquired of them that be drowned, or suddenly dead, and after it is to be seen of such Bodies, whether they were so drowned or slain, or strangled by the Sign of a Cord tied straight about their

Necks, or about any of their Members, or upon any other Hurt found upon their Bodies.[4]

Inquests were held before a jury of between 12 and 15 men. By the nineteenth century it was incumbent on the newly created Irish constabulary to notify the local coroner when a dead body had been discovered. The coroner decided whether or not an inquest was necessary. Legislation introduced in 1836 made provision for the attendance and remuneration of medical witnesses at coroners' inquests and, while doctors did not attend all inquests, it was unusual if one or more did not testify.[5]

The majority of suicides reported in nineteenth-century Ireland were judged temporarily insane at inquests.[6] This meant that no crime had been committed. Was such a verdict merely a legal loophole or was the link between suicide and insanity firmly established within wider circles of Victorian Ireland? After 1872 legal punishment no longer existed but religious prohibition remained. However, the Protestant and Catholic churches in Ireland accepted the legal verdict of temporary insanity as a sign that no sin had been committed and that the suicide could be buried in consecrated ground. When James MacCullagh, an eminent mathematician at Trinity College, Dublin, died by his own hand in October 1847, the *Freeman's Journal* reported that he was conveyed in a horse-drawn hearse to the church in Strabane. His body was laid in the church for the night prior to his burial and the ceremony began at ten the following morning. 'The shops and places of business were closed and the gloom which pervaded all classes exhibited most forcibly in the extreme veneration in which the deceased was held.'[7] Had MacCullagh been found *felo de se* he should, according to the law, have been buried at night under police escort.[8] The Catholic Church had a similar understanding of the role of mental illness in suicide and followed legal tradition by using this reason to excuse the sin of self-murder. In 1860 a 19 year old student Thomas Maginn died by his own hand while studying for the priesthood at Maynooth College, the centre of the Catholic hierarchy in Ireland at the time. He was buried in consecrated ground.[9] Before he died Maginn made a full confession and this would also have enabled him to receive ecclesiastical burial. The verdict of temporary insanity, supported by a senior college staff member, also permitted his burial in the college cemetery. Recording the verdict in his diary Bishop McCarthy stated 'the reason was entirely gone when [Maginn] ... admitted these suggestions on the part of the devil. It is unnecessary to add the verdict – died by his own hands, being of unsound mind at the time.'[10]

Many suicides would not have been able to make a full confession between inflicting the lethal wound and dying, but in such cases the Catholic Church followed the legal presumption of 'innocent until proven guilty'. Sanity rather than insanity had to be proven at inquest and

> In default, then, of proofs or of their certainty, and even in case of doubt as to the deliberation of the act and the nature of the motives for committing suicide, the decision taken must always be in favour of the deceased, who is presumed to have taken his life in a moment of mental aberration, and while irresponsible for his acts ... and therefore he could not under the circumstances, be punished and deprived of ecclesiastical burial.[11]

In a country where religion was of such importance and where both the Protestant and Catholic churches had experienced rejuvenation in the post-Famine period, there was an incentive to ensure that a verdict of temporary insanity was returned at inquests even after legal punishment had been removed from the statute books.[12] Durkheim's seminal work on suicide attributed its rise in Western Europe and America in the nineteenth century to a decline in religious faith and increasing secularisation.[13] An article on the increase of suicide published in the *Journal of Mental Science* in 1899 stated that religion influenced the frequency of suicide and that it was rare in Ireland because 'the Irish, whether Catholic or Protestant, are fervid in their religious views'.[14] However, suicide in Ireland rose during this period, even after the rise in religious practice and observance within both churches.

Chart 4.1   Suicide in Ireland per 100,000, 1864–1919

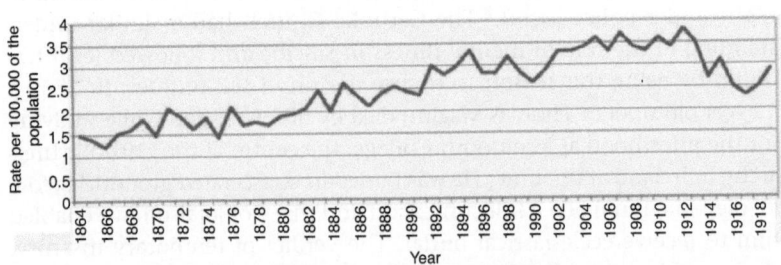

Source: *Annual Reports of the Registrar General, 1864–1920*.

Chart 4.1 above shows the rising rate of suicide in Ireland between 1864 and 1919. The figures are taken from the annual report of the Registrar General, an office established in Ireland in 1863, 25 years after the creation of a similar office for England and Wales.[15] They reveal a rise in the *reported* rate of suicide although it should be noted that there are numerous problems with these figures.

During this period records show that it was difficult for the coroner to hold an inquest in some cases, and this had an impact on the accuracy of the figures. For example, in 1879 the coroner for Mayo stated that in his district (Ballinrobe and Westport) eight bodies had been buried before he could get to the place where they had died and he 'did not like for many reasons to have them exhumed'.[16] He recommended that the Royal Irish Constabulary be allowed to use the telegraph service to inform the coroner of the need for an inquest. This would save the constable from having to walk two days to inform the coroner and, as he noted, 'it seems absurd to put such an amount of labour on the men for the saving of a shilling'.[17] In 1878 there were eight cases of sudden death in which the body of the deceased had been buried by the time the coroner arrived in the area. The legal process was prevented by this delay in notifying the coroner and the burial of the body within three days. Administrative problems, finance, and religious custom occasionally undermined the legal investigation into sudden death.

Even when coroners did hold inquests, accuracy in the recording and registering of suicidal deaths was not guaranteed. In county Clare coroner John Lynch presided over two cases in the early twentieth century where suicide notes had been left, outlining the actions of the deceased and giving information on the method (drowning) and the probable location of the bodies. At these inquests into the deaths of Susan McHugh in 1913 and Henry Cullinane in 1918, the juries returned verdicts of 'found drowned'.[18] These deaths would not have been recorded as suicides by the registrar, who would have counted them as 'found drowned' or 'open verdict' in his annual reports. Despite the compelling evidence of a note, there was no mention of suicide in the verdicts.

The death of landlord Thomas Judkin Fitzgerald in April 1864 provides a possible explanation for these surprising verdicts from county Clare and helps us to understand more fully the importance of the link between suicide and insanity and its impact on the accuracy of official statistics. Fitzgerald drowned himself in a river near his house in Golden, county Tipperary having written two letters to family friends instructing them on what to do once they heard the bad news. Before an inquest was held and the contents of these letters were revealed, the earliest newspaper report of his death suggested that he died as a result of 'an act committed, we are inclined to think, while the reason was temporarily unseated'.[19] For readers of nineteenth-century newspapers this statement implied that he died by suicide but the assumption that he was insane proved controversial. The coroner, when instructing the jury on their duty to bring in a verdict, told them

> If you believe that, at the time of committing this act, he was of unsound mind, then your duty will be to find that he committed suicide during some interval of insanity. If you think he was sane at the time he committed the deed, it is your duty to bring in an open verdict of found drowned.[20]

This instruction suggests that suicide was concealed behind verdicts of 'found drowned' in cases where a jury did not believe the deceased was insane at the time of death. Local magistrate, the Honourable M.J. ffrench challenged the coroner on this instruction, stating that 'we admit the imputation that myself or any other person had murdered him'. The foreman of the jury stated that 'an open verdict is not an imputation that any person in the county murdered him'. Massey, one of the witnesses and a justice of the peace questioned the coroner, 'would it not be well to put it to the jury ... and I ask you now, from your own knowledge as a medical gentleman, that they are to consider whether in point of fact, that the very act of taking life in itself establishes insanity'.[21] Ffrench responded by saying that 'if the jury believed the deceased to have been in his senses, that a verdict of found drowned would not be proper one'; in other words, he should have been found *felo de se*. The jury's verdict stated 'That the said Sir Thomas Judkin Fitzgerald did wilfully drown himself in the river Suir, on the morning of the 27th of April, 1864, while labouring under temporary insanity.'[22]

The majority of suicide verdicts passed in Ireland, and those that were registered as suicides, were returned 'temporarily insane'. The evidence discussed above suggests that those suicides considered sane could be concealed under open verdicts. Legal punishment before 1872 and the continued existence of religious punishment for sane suicides, in the form of burial outside consecrated ground, provided an incentive for such concealment. This suggests that the verdict of temporary insanity could be understood as a loophole that prevented punishment and the link between suicide, insanity and medicine could be seen as merely a construct to evade legal and ecclesiastical authority.

Treatment of suicidal behaviour: Asylums and prisons

In order to examine the treatment of suicidal behaviour we must first understand what was considered suicidal. Contemporary explanations for suicide given in the posthumous circumstances of an inquest are problematic because of the influence of hindsight. Under questioning from police, coroners and others, witnesses were prompted to examine behaviour and incidents that, although they were not thought odd at the time, probably became significant in light of a sudden and violent death. At inquests there were many different types of behaviour which, when presented, pointed to a possible mental illness. At the inquest of David Cathcart in Fermanagh in 1919 his daughter-in-law described him as having a 'wild stare in his eyes' on the day he shot himself.[23] The doctor thought that Cathcart's recent bout of influenza had 'unhinged his brain power'.[24] In another case, William Christy Jnr of Belfast thought his father, William Christy Snr, had been 'restless and drowsy' for the two months before he drowned himself in November 1891.[25] Although physical illness was mentioned in the histories of both of these men, there was no sign that they received medical treatment for suicidal feelings or mental illness. So did families seek medical assistance for those who were considered suicidal before they died?

Sources show that this seemed to vary from case to case. When Sarah Campbell of county Down, a 47 year old farmer's wife, shot herself in 1921, her husband revealed that she had threatened to do away with herself only the day before. He described how she believed herself to be suffering from consumption, a fact confirmed by the dispensary doctor she had consulted 'in May, June and July'. The doctor stated that her fears about lung problems were 'imagination'. Despite this 'imagination', she received 'no special treatment' and even after her threat of suicide no doctor was called.[26] The doctor at the inquest of John Keating in August 1888 stated that the deceased's daughter had told him 'quite lately he said that he would make away with himself'.[27] The previous month John McGlynn's housekeeper told the coroner that she 'heard him saying he would put a hand to himself if he did not soon get better'. She 'spoke to him about it and tried to put him off and that was all there was about it'. He had been suffering from 'general debility [for twelve months] and lately … sleeplessness'.[28] These cases suggest that medical advice was not sought for those who threatened suicide but evidence elsewhere reveals that this was not always the case.

In some cases suicidal people were cared for at home. James Farrell of Newry began 'talking to himself', saying he 'could not live' after the death of his wife in 1921. The doctor who attended him 'allowed ... [Farrell] ... to remain at home on the terms that he would be looked after by a responsible person'.[29] In the main, however, it does appear that those who expressed a wish to die were often referred to a doctor and medical assistance through the asylum system, as either public or private patients. In 1912 Lord Castletown of Doneraile, county Cork, threatened suicide. His wife and his cousin, Lord Cloncurry, had him confined to a private asylum at St Edmondsbury in Lucan shortly this threat.[30] Castletown's confinement in Lucan did not last very long; he was released within a couple of months to the care of his sister and went on to live until 1937, dying a natural death. An Act of 1842 governing private asylums provided that 'all admissions should be at the instance of an order made out by a relative or friend'.[31] These admissions also required medical certification from two doctors who had examined the patient. As Finnane points out, the innovations in the Irish asylum system in the 1840s gave 'medical men a pre-eminent role'.[32]

Castletown's confinement in the private asylum was certainly the result of his social position but the decision to have him admitted was not peculiar to the wealthier classes. The mother of Bridget Geoghan, who drowned herself in July 1899 in county Clare, revealed at the inquest that Bridget had attempted suicide the previous year. Her mother had tried

> to get her committed to the [district] asylum. About the 9th of last June she attempted to drown herself ... the magistrates refused to commit her and she was in the County Infirmary for a short time. She was better since she came home.[33]

The dispensary doctor George McNamara testified that he had attended her on several occasions. She 'was suffering from melancholia with suicidal tendencies. I inferred from what her mother told me her head was wrong'.[34] Geoghan's mother mentioned the magistrates' refusal to commit her daughter and, here, she was referring to the Dangerous Lunatics Act of 1838 (amended in 1867), which meant that those considered 'dangerous' either to themselves or others, could be committed first to gaols and, after the 1867 amendment, directly to district asylums once two magistrates signed the committal form.

The 1867 amendment also required that a medical certificate be signed by a dispensary doctor from the district in which the magistrates were sitting.[35] According to Finnane, this was the 'most important mode of admission' but a suicide attempt itself was not 'necessary; threats to do away with oneself were commonly cited as indicating the need for asylum care'.[36] A significant number of inmates were classified as suicidal on admission to public asylums. Although in 1874, 485 district asylum patients were described as suicidal (just under 3 per cent of the total), figures for the year 1880 reveal that no one was confined in an asylum under the Dangerous Lunatics Act for attempting suicide. This changed by 1890, however, when a total of 277 were confined.[37]

During this period, as admissions to the asylums for suicidal individuals were rising, prosecutions for those who attempted suicide were also on the increase. Attempted suicide was a criminal offence in Victorian and Edwardian Ireland, a relatively 'new' crime that emerged during the mid-Victorian period at the same time as the professional police expanded and their concerns with public order, welfare and decency became prominent.[38] The numbers of those proceeded against varied between 11.9 per cent in 1895 and 50 per cent in 1915. Where proceedings were initiated against individuals the majority resulted in convictions; 47 per cent were convicted in 1896 and as many as 85 per cent were convicted in 1908. This implies that the state did not proceed against attempted suicides unless they had a strong case, or strong evidence against the individual.[39]

In cases where the state did proceed against individuals, they did not always imprison them. Judicial statistics show that, as a proportion of the crimes known to police, those who were convicted and imprisoned were very low.[40] In 1895 nine of the 13 people convicted of attempted suicide were released on recognizance.[41] Maud O'Mahony, found insane on indictment in 1897, was sent to an asylum but she was later released to her friends on the recommendation of asylum authorities.[42] Between 1895 and 1899 nine people were either found guilty but insane or were judged insane on arraignment. During those years 82 people were convicted of attempted suicide.[43] The prime motivation behind the imprisonment of attempted suicides appears to rest on the potential for observation and surveillance in prisons. In 1916 the Recorder in Belfast stated that he only passed sentence following consultation with Dr Fulham:

> The sentence imposed was more with a view to the prisoner's safety and improvement in health, physical and mental, than to punishment. I would regard his further detention in the same light and to be decided by a medical expert.[44]

In cases of suicide attempts, judges used the option of imprisonment not as a means of punishment but to secure the individual's safety. This reflects what McGowen has said about English prisons in the nineteenth century: 'as prisoners were more closely controlled and observed, punishment disappeared from the experience of most people'.[45] Workhouses similarly took care of those they deemed unfit to care for themselves and, when paupers attempted suicide, guardians usually tried for committal into the local asylum, rather than trying to have the individual prosecuted. In 1861 the Guardians of North Dublin Union directed the master to 'try for admission into [the] lunatic asylum' when Patrick Clarke made a wound in his throat.[46]

Conclusion

This paper explored legal, religious and popular links between suicide and insanity in order to determine the nature of that relationship in Victorian and Edwardian Ireland. Evidence suggests that it was not merely a remnant of times past when a verdict of temporary insanity at an inquest provided protection from legal and ecclesiastical punishment. The testimony heard at inquests, official figures from asylums relating to admission, and decisions made by judges following attempted suicide, suggest that Victorian Irish society was concerned with the safety and surveillance of suicidal individuals. In many cases family and friends chose to have their relatives cared for medically, in their homes by local doctors or in public and private asylums. Judges who made decisions about custodial sentences were concerned for the health and welfare of those who were convicted of attempted suicide.

In the period after 1872 when legal punishments for suicide had dissipated, the attitude of the churches to sane suicides remained a motive for ensuring the survival of a verdict of temporary insanity. However, records show that those within the churches, both Catholic and Protestant, believed that all suicides should be assumed insane rather than sane, offering the benefit of the doubt to the deceased and their families. Nevertheless, examples from the 1860s and 1910s suggest that even in cases where suicide notes existed, an attempt was made to conceal the suicide officially. This suggests that concealment of suicide took place only when the individual appeared sane. However, the majority of those few cases in which suicide notes survive were found insane, thereby challenging the idea that concealment of sane suicides was a widespread practice. The relationship between suicide and insanity does appear well established in Ireland by the post-Famine period in legal, religious and popular circles. However, while some families sought medical advice and treatment when faced with someone who appeared depressed or suicidal, others, like the families of Sarah Campbell and John McGlynn, did not.

There was no uniform reaction to suicide in Ireland at the time, but medical interpretations appear to have been significant as a paradigm used to explain and explore self-destruction.

Notes

1 For information on the increasing medicalisation of both suicide and insanity see Michael McDonald, *Mystical Bedlam: Madness, Anxiety and Healing in Seventeenth-Century England* (Cambridge 1981); Michael MacDonald and Terence R. Murphy, *Sleepless Souls: Suicide in Early Modern England* (New York, 1990); Andrew Scull, *The Most Solitary of Afflictions: Madness and Society in Britain, 1700–1900* (New Haven and London, 1993). For a discussion of the increasing role of medical professionals in the newly established state-sponsored lunatic asylums in Ireland see Mark Finnane, *Insanity and the Insane in Post-Famine Ireland* (London, 1981), especially pp. 39–47. No work has been published on the history of suicide in Ireland except for Mark Finnane's 'A decline in violence in Ireland: Crime, policing and social relations, 1860–1914', in *Crime, History and Society*, 1 (1997), pp. 51–70.

2 The moveable property of felons, including suicides, was forfeit to the crown once the suicide was found guilty or *felo de se*. Clodagh Tait, *Death, Burial and Commemoration in Ireland 1550–1650* (New York, 2002), p. 22. This was revoked in 1872 with the passage of 33 & 34 Vict.c.23 Forfeiture Act. Since 1823 ignominious burial of suicides found *felo de se* was outlawed under 4 Geo.IV.c.52 (U.K.) Felo de se Act. After that, suicides had to be buried at night under police escort. This law was revoked in England in 1880 and 1882. In Ireland the Catholic Church and the Church of Ireland retained authority regarding burial of suicides after Disestablishment of the Church of Ireland. 4 Geo.IV.c.52 (U.K.) Felo de se Act; 33 & 34 Vict.c.23 (U.K. except Scotland) Forfeiture Act.

3 Brian Farrell, *Coroners: Practice and Procedure* (Dublin, 2000), p. 5; 10 Hen.7.c.22.9.

4 Matthew Dutton, *The Office and Authority of Sheriffs, Under-Sheriffs, Deputies, County Clerks and coroners in Ireland* (Dublin, 1721), p. 6.

5 Coroners Act, 1836, 6 & 7 Wm.c.89.

6 This is based on a random sample of almost 1,000 suicides that occurred between 1831 and 1921 from inquests, government reports, newspapers and elsewhere. See *Return of the numbers of inquests held by the coroners of the counties and cities in Ireland, 1841–1842* (206), xxxviii, p. 185.

7 *Freeman's Journal*, 8 November 1847. See also Raymond Flood, 'Mathematics in Victorian Ireland', *British Journal for the History of Mathematics*, 21 (2006), pp. 200–11.

8 Geo.IV.c.52 (U.K.) Felo de se Act.

9 John Healy, *Maynooth College, 1795–1895* (1895), Appendix XVIII, p. 739. An earlier suicide occurred in February 1841. The student in that case, John O'Grady from Limerick, was also found insane and was buried within the general cemetery population in the college grounds. See Georgina Laragy, 'Suicide in Ireland, 1831–1921: A social and cultural history' (unpublished PhD: NUI Maynooth, 2005), pp. i–ix.

10   Michael Ledwith (ed.), *The Diary of Dr Daniel McCarthy* (Maynooth, 1984), 21 April 1860.
11   'Notes and Queries', *Irish Ecclesiastical Record*, 23 (January–June 1908), p. 99.
12   Emmet Larkin, *The Consolidation of the Roman Catholic Church in Ireland, 1860–1870* (Chapel Hill, 1987); Janice Holmes, *Religious Revivals in Britain and Ireland 1859–1905* (Dublin, 2000).
13   Emile Durkheim, *Suicide: A Study in Sociology*, trans. by John A. Spaulding and George Simpson (New York, 2000).
14   William Ireland, 'On the causes of the increase of suicide', *Journal of Mental Science* (July 1899), p. 461.
15   General Registry Office, *Registering the People: 150 years of Civil Registration* (Dublin 1995).
16   Criminal and Judicial Statistics of Ireland, CSO Part One, Official Papers, 1878/17, National Archives of Ireland (NAI). Hand-written note on the end of the return made by James Rutledge, coroner for Ballinrobe and Westport District, county Mayo, dated 1 February 1879. Although returns were obviously made for 1878 they were not included in the printed British parliamentary papers. These returns exist in manuscript form in the NAI.
17   Ibid.
18   Clare Inquests; Susan McHugh (1913) ID.33.173 and Henry Cullinan (1918) ID.39.113. NAI.
19   *Clonmel Chronicle*, 27 April 1864.
20   *Tipperary Express and Advertiser*, 30 April 1864.
21   Ibid. A similar report of the inquest including the information above appeared in the *Tipperary Free Press and Clonmel General Advertiser*, 29 April 1864.
22   *Tipperary Express and Advertiser*, 30 April 1864.
23   Coroners Inquests, Fermanagh, Fer 6/1/1/5. Public Record Office of Northern Ireland (PRONI).
24   Ibid.
25   Coroners Inquests, Antrim, Ant 6/1/1/2. PRONI.
26   Down Inquests, Down, Dow 6/1/1/7. PRONI.
27   Armagh Inquests, Armagh, Arm 6/1/1/1. PRONI.
28   Armagh Inquests, Armagh, Arm 6/1/1/1. PRONI.
29   Down Inquests, Down, Dow 6/1/1/7. PRONI.
30   Doneraile papers, Ms.34,166 (3–4). Letters from Lord Cloncurry to Lord Castletown, June–September 1912. NLI. Archives of St Patrick's Hospital, Dublin, Register of Admissions, St Patrick's Hospital, E/108, Patient Number 86. Both Lord Cloncurry and Lady Castletown had personal experience of suicide. Cloncurry's father committed suicide in 1856 and his two sisters were also believed to have committed suicide. James M. Cahalan, 'Forging a tradition: Emily Lawless and the Irish literary canon', in Kathryn Kirkpatrick (ed.), *Border Crossings: Irish Women Writers and National Identities* (Dublin, 2000), p. 40; Elizabeth Grubgeld, 'Emily Lawless's *Grania: The Story of an Island (1892)*', *Éire-Ireland: A Journal of Irish Studies*, 23:3 (1987), p. 124. Godfrey Levinge, land agent at the Doneraile estate, killed himself in the 1890s. See NLI, Doneraile papers, Ms.34,033 (8–9) for information on Levinge's death.
31   Finnane, *Insanity and the Insane*, p. 91.
32   Ibid.

33 Clare Inquests, 1851–97, I.D. 30.112. NAI.
34 Ibid.
35 Finnane, *Insanity and the Insane*, pp. 90, 96–7.
36 Ibid., pp. 91, 151.
37 House of Commons Parliamentary papers, *Twenty-fourth Report on District, Local and Private Lunatic Asylums in Ireland*, 1875 [C.1293], p. 147. This is a proportion of the total number of patients in the asylums, 1770–5. Hereafter these reports from asylums will be listed as asylum reports. *Thirtieth Asylum Report* (1881) [C.2933], xlvii, p. 469; *Fiftieth Asylum Report* (1901) [Cd.760], xxviii, p. 487.
38 Olive Anderson, *Suicide in Victorian and Edwardian England* (Oxford, 1987), p. 423.
39 Figures are based on the criminal and judicial statistics published annually by the government. 1897 *Return of Judicial Statistics of Ireland, 1895* (Part I. Police; Criminal Proceedings; Prisons) [C.8616]; 1917–18, *Return of Judicial Statistics of Ireland, 1915*, xxiii, p. 8636. See House of Commons Parliamentary papers online.
40 Between 1863 and 1875 only 12 per cent of known attempted suicides were convicted. *Judicial Statistics*, 1864–1876.
41 *Judicial Statistics*, 1897 (C.8616).
42 Convict Reference Files (C.R.F.), 1897/O/14. NAI.
43 *Judicial Statistics*, 1897 (C.8616). This does not include figures for 1897 as they are not currently available either online or in the National Library of Ireland.
44 C.R.F., 1916/H/11. NAI.
45 N. Morris and D.J. Rothman (eds), *The Oxford History of the Prison: The Practice of Punishment in Western Society* (Oxford, 1995), p. 98.
46 BG/A/1-114. North Dublin Union, Board of Guardians, Minutes, 30 January 1861. NAI.

5
Psychiatry and the Fate of Women Who Killed Infants and Young Children, 1850–1900

Pauline M. Prior

Introduction

Statistics on crime across time and place point to the fact that women are much less likely than men to engage in criminal behaviour.[1] However, there is one area of crime where women are highly visible: that of child killing. A number of terms are used to describe this crime, each representing an attempt to define the age of the victim. The most common terms are 'child killing', 'baby killing' and 'infanticide'. Child killing is used to describe the overall picture with victims ranging in age from new-born to adolescence, though the most vulnerable age is immediately after birth; in other words, the killing of older children is a rare crime.[2] The term 'infanticide' is used to refer to crimes in which the victim is under the age of 12 months. However, though the phrase was used in medical and official correspondence in nineteenth-century Ireland, the *crime* of infanticide did not exist during the nineteenth century.[3] In Ireland, the Infanticide Act was not passed until 1949, while in England it came on the statute books in 1922 and was updated in 1938.[4]

Before the infanticide law was passed, women who killed a child (even if this child was a newly born infant) were faced with the prospect of being found guilty of either murder or manslaughter and sentenced to death or imprisonment, unless it could be shown that they were 'insane' at the time of the crime. These women were generally sent to the Central Criminal Lunatic Asylum at Dundrum (hereafter, Dundrum Asylum), county Dublin, as either 'unfit to plead' or as 'guilty but insane' at the time of the crime.[5] By the early twentieth century, there was a growing awareness of the need for 'infanticide' legislation, to separate these crimes from ordinary murder and manslaughter but, until it appeared on the statute books, the most common legal approach was the use of Section 60 of the Offences Against the Persons Act (24 & 25 Vic. c. 100), which allowed for the lesser conviction of 'concealment of birth', if the child was newly born.[6] In this discussion, the focus is on women who came before the Irish courts for the killing of one or more children, ranging in age from new-born to five years old, in the second half of the nineteenth century.[7] Unlike many other areas of crime, this was one in which psychiatrists played an important role in determining the outcome of the court proceedings. When sought, a psychiatric opinion on the mental state of the mother had a profound effect on what happened to her.

The period 1850 to 1900 in Ireland was characterised by changes in many aspects of social and economic life. These were brought on by the decline in population following the Great Famine of the 1840s, the consolidation of British rule as reflected in the expansion of institutional responses to law and order problems, and the growing political discontent of large sections of the population.[8] All of these changes had an impact on family life in general and on the lives of women and children in particular. In the discussion that follows, we will see how economic hardship, combined with social stigma, could lead to a situation of extreme violence between a mother and her young child. In the case of illegitimacy, this violence was often seen as the solution to the social and economic problems caused by the child's birth. However, then as now, this was not the end of the story. Child killing by a mother, when it became public knowledge, was regarded as one of the most incomprehensible of actions. In the words of Dr F.X. MacCabe, Medical Officer of the Prisons Board in 1886, it was 'a perversion of the maternal instinct'.[9] It required a public response that allowed for some form of severe punishment for those deemed fully responsible for their actions and an element of clemency for those who were not.

Killing a young child

In Ireland, the societal response to the killing of a young child by his or her mother was quite complex. This is an under-researched area of Irish history, but existing studies suggest that, when compared with England, Ireland had lower rates of both illegitimacy and of infanticide during the nineteenth century.[10] However, according to McLoughlin, friends and family members of the mother often went to great lengths to conceal the birth and death of an unwanted child, which led to a low rate of reporting of the crime to the authorities.[11] Because of this under-reporting, it is very difficult to be certain of the extent of infanticide or of the killing of older children in Ireland. The best estimates of child killing rates are those compiled by O'Donnell from police records and judicial statistics.[12] He suggests that the highest level of 'baby killing' (victims under one year) was during the period 1851 to 60, an average annual rate of 18.3 per million of the population. This represented 49 per cent of the average annual rate for all homicides for that period. This pattern is consistent with, though lower than, the pattern for England in the mid-nineteenth century, where infanticide (victims under one year) accounted for 61 per cent of all homicide victims.[13] O'Donnell also found that, in Ireland, the officially reported rate of infanticide declined rapidly after the famine, as did the overall homicide rate, so that in the years 1891 to 1900, the average annual rate for 'baby killing' was 4.4 per million of the population, representing 19 per cent of the average annual rate for all homicides during that period.[14] Debates on the reasons for the decline in this type of crime include arguments on the influence of the famine and of religious beliefs about illegitimacy (especially considering the powerful influence of the Catholic Church), the impact of improvements in child health, and changing attitudes to children within Irish society.[15]

An exploration of the records of women who appeared before the Irish courts for the killing of an infant or a young child during the second half of the nineteenth century, shows that they could be dealt with in a number of ways: acquittal through lack of evidence; conviction of murder or manslaughter, with the concomitant sentences of execution or imprisonment; a verdict of 'guilty but insane', leading to an indeterminate sentence in the Dundrum Asylum; or, after 1861 and for those who had recently given birth, the lesser conviction of concealment of birth.[16] The use of this 1861 legislation, which led to a much lighter sentence, was encouraged by the Home Office in London, which by then had a fairly compassionate attitude towards infanticide.[17]

According to convict records, 19 women were sentenced to death for child murder (including some who had killed new-born infants) in the period 1864–1902, but these sentences were never carried out.[18] The only three women to be executed in Ireland during the second half of the nineteenth century were all involved in the murder of adult men.[19] In situations where women were found guilty of child murder, the death sentence was invariably commuted to penal servitude for life or for a specific period of time. Roger Smith, in his study of nineteenth-century trials in England found similar patterns of sentencing for child-killing, including infanticide. For example, he estimated that between 1849 and 1864 there were 39 convictions for child murder by a mother, some of whose victims were newly born infants, but that very few were executed for the crime. In fact, the last woman hanged for infanticide in England was Rebecca Smith in 1849.[20] Her crime was much more extreme than those that normally came before the courts; she had admitted to killing seven of her 11 children with rat poison.[21]

Women sent to Dundrum Asylum were designated as 'criminal lunatics'. Research by Mulryan and colleagues on statistics from Dundrum reveals that from the time of its opening in 1850 to 2000, 37 women were confined there for killing children, mostly children under one year old.[22] These women had been judged by the courts as either 'guilty but insane' or 'unfit to plead', verdicts that were based on reports from psychiatrists called to assess the women after arrest. This assessment only took place if a woman exhibited symptoms consistent with what was then regarded as insanity. Before discussing some of these cases in more detail, we shall look first at the context within which Dundrum Asylum functioned.

From the beginning of the nineteenth century in Ireland, vagrancy and lunacy were subjected to waves of legislation to ensure control of unruly elements in society. This was due partly to the enthusiasm for the reform of the Irish asylum system of two Whig politicians, Thomas Spring-Rice and Sir John Newport, and partly to the ease with which the highly centralised colonial authority exerted control over local authorities and landowners.[23] As the general population decreased, from approximately eight million in 1841 to 6.5 million after the famine and to 4.5 million in 1900, the prison system and the asylum system expanded steadily.[24] The case of the prison system is interesting because scholars debate whether or not serious crime actually increased in the second half of the nineteenth century.[25] As part of the planned expansion of the asylum system and to solve the problem of lunatics who were already in an overcrowded prison system, a special facility, with accommodation for approximately 100 people, was built at Dundrum, Dublin, for the confinement and treatment of criminal lunatics. Opened in 1850, it pre-dated the establishment of a similar institution at Broadmoor, England (1863), but it mirrored the ideological approach to crime and mental disorder evident throughout Britain and its colonies at this time.[26] The Dundrum-based institution was known as the Central Criminal Lunatic Asylum for Ireland and remains open today as the Central Mental Hospital for offenders with mental disorders.

The plans for the proposed institution were influenced by the largely benevolent and optimistic perspective of Dr Francis White, the Inspector of Asylums for Ireland. Because he saw the potential inmates as 'lunatics', rather than 'criminals', he argued for a building that would be more like an asylum than a prison. In his annual report of 1847 he wrote: 'It is not designed that the building should partake of the character of a prison' and 'it is proposed to have the structural arrangement as cheerful as circumstances will admit, so as to afford every possible facility for the recreation and occupation of the patients'.[27] In spite of this positive approach, Dundrum Asylum became a place to be feared. Individuals confined there were regarded by the public as both dangerous and disordered and were isolated from society by virtue of physical barriers, such as high periphery walls, and legal outcomes, such as indefinite sentences.

Women sent to Dundrum Asylum for killing a child

An exploration of medical records at Dundrum Asylum, from its establishment in 1850 to the end of the century, reveals that one of the clear patterns of crime to emerge in relation to female patients is the frequency of child killing among those sent there for involvement in a murder. Tables 5.1 and 5.2 give a brief outline of some of the social characteristics of these women: their age, marital status and diagnosis.

Table 5.1   Women who killed an infant (age under one year) and found to be insane at the time of the crime

Name	Age	Marital status	Diagnosis	In Asylum*
Mary Glass	21	Single	Congenital idiot	1868–1902
Kate Connor	18	Single	Chronic mania	1888–1902
Mary Finnegan	31	Married	Chronic mania	1889
Agnes Rennie	31	Married	Melancholia	1889
Margaret Rainey	19	Single	Dementia	1891–94
Mary O'Flaherty	34	Married	Melancholia	1892
Ellen Byrne	26	Single	Puerperal insanity	1893–1894
Hannah Sullivan	17	Single	Dementia	1895–1896
Ann McDonnell	38	Married	Melancholia	1900–1910

*Some went to Dundrum Asylum and some to a local District Asylum
Sources: Medical records (Dundrum Central Mental Hospital), Royal Irish Constabulary (RIC) and Convict Reference Files (National Archives of Ireland).

Table 5.2 Women who killed a child (age over one year) and found to be insane at the time of the crime

Name	Age	Marital status	Diagnosis	In Asylum*
Johanna Doyle	40	Married	Chronic mania	1888–1895
Sarah McAlister	33	Married	Melancholia	1892
Catherine Wynn	35	Married	Melancholia	1893–1911
Mary Jane Simpson	45	Married	Melancholia	1893–1901
Mrs Sadlier	30s	Married	Mania	1896

* All, except Ellen Sadlier were in Dundrum Asylum
Sources: Medical records (Dundrum Central Mental Hospital), RIC and Convict Reference Files (National Archives of Ireland).

Some of the information given here was verified in a number of sources, whereas other women featured in only one source. This has to do with the delicate nature of an insanity conviction, and the different methods of record-keeping in the prison system and the asylum system. Because all of the patients in Dundrum Asylum had been indicted for a crime, their names were (and still are) in the public domain. Their cases were reported fully in court records and some made it to the newspapers. However, few case notes for patients in Dundrum have survived, though there was an obligation on the administration to keep individual files on patients, as there was for the prison system regarding convicts. The best source of information is often the convict record file, containing correspondence that took place between, or on behalf of, an individual patient and the office of the chief secretary for Ireland, which handled all the appeals for early discharge. However, unlike ordinary convicts, these asylum patients could be discharged on the recommendation of the medical superintendent of Dundrum Asylum, backed up by the inspectors of lunacy, without any intervention on the part of the patient or his or her family. In some cases, this was a very short procedure, leaving a small 'footprint' for the researcher.

Existing records suggest that women who killed a new-born child during a period of temporary insanity, and who regained sanity quickly in Dundrum Asylum, were recommended fairly quickly for discharge. Each case had, of course, to go before the lord lieutenant but, as outlined in the report of the inspectors of lunacy in 1866, great 'clemency' was shown by his office. The inspectors wrote that, in England, an acquittal of homicide on the grounds of insanity usually led to a sentence of life-imprisonment, whereas in Ireland, the lord lieutenant had 'several times extended his clemency to such persons' and 'no evil consequences' had emerged from their early release. They gave the example of a young mother who 'in a paroxysm of puerperal mania, destroy[ed] her infant', but who recovered quickly, realising her crime and her loss. They ask 'is it not better if she goes to her family'?[28]

Women who did not recover their sanity were kept in Dundrum Asylum if their behaviour was difficult to control, but they were discharged to their local district asylum if they became calm and easily managed. Table 5.1 shows information on nine women who were regarded as infanticide cases: women who had killed a baby under the age of 12 months. Margaret Rainey, for instance, was a 19 year old Episcopalian servant from Belfast who, in 1891, was found 'not of sound mind or capable of pleading' in relation to the murder of her illegitimate daughter. She had delivered the baby safely, with the doctor present, in lodgings that her sister, Jane, had arranged on her behalf. Jane had also arranged for the baby to be sent to a nurse soon after its birth, so that Margaret could take up a position 'in service' without the burden of a child. However, Margaret was not happy with this arrangement. She would have preferred to keep the baby at home, but Margaret's mother refused to allow her home with the child. On the night of the crime, Margaret had been left alone with the baby for a short period in her lodgings. At around eleven o'clock that night, she alerted a neighbour, saying that her baby had been 'stolen by a person or a dog'. Sadly, the baby was found dead in the yard next door. The medical opinion was that it had died as a result of injuries sustained from a fall, probably from an upstairs window. Margaret denied having thrown the baby through the window and insisted that it had been stolen. After being assessed as 'not of sound mind' by a doctor in her local area, she was sent to Dundrum Asylum where her initial diagnosis was dementia. This diagnosis changed over time, so that two years later in 1893, she was assessed by Dr Revington, resident medical superintendent, as 'weak-minded rather than insane'. She was discharged in 1894 to her sister Jane, who had supported her throughout her ordeal.[29] Margaret probably had what is now considered to be an intellectual disability. She was lucky in that she did not have to appear in court and that she spent only two years in confinement for a crime that could have carried the death sentence.

Another young single woman in a similar predicament was Hannah Sullivan, a 17 year old Catholic servant from county Cork. She was indicted for the murder of her new-born child in 1895 and was found 'guilty but insane' by the court. According to Dundrum Asylum medical records, she killed the baby 'by cutting off its head in a loft of her master's premises at Tralee'.[30] Hannah herself said that it was an accident, that she did not know what was happening, as she did not realise she was expecting a child. She also said that the baby's head had got caught in the toilet seat as she gave birth. Hannah may have been telling the truth when she said that she did not realise that she was giving birth, as this was a story repeated by other young unmarried women whose babies died in suspicious circumstances. In all of these cases, the young women were alone at the time of the birth.

Similar stories were told by young English women who were tried for new-born child murder. Margaret Arnot writes of Elizabeth Cornwall, a 38 year old servant from London, who delivered her baby in a 'chamber pot' in her employer's house in 1847.[31] Elizabeth left the baby in the pot until the next day, when she emptied it into a 'slop pail' and threw it into the 'water closet'. Elizabeth's crime was discovered only when she told a friend about her actions and the distress she was feeling. When she found herself accused of murder before the court, she told the judge, 'I never saw it, as it came from me in the chamber utensil, so it remained'.[32] She also said that she did not know she had been pregnant. While this admittedly seems an unlikely explanation for someone of Elizabeth's age – she was 38 at the time – it was perhaps more believable in the case of the 17 year old Hannah Sullivan. As Arnot suggests, the denial of a pregnancy was one way of avoiding the responsibility of deciding what to do after giving birth, an event that would destroy the character and life chances of women such as Elizabeth and Hannah. Luckily for Hannah, she was treated with compassion in Dundrum Asylum and, after showing no signs of insanity, she was discharged to her mother after one year in detention.[33]

The official view of the medical staff involved in the forensic psychiatric services of the time is reflected in the annual reports on Dundrum Asylum by the inspectors of lunacy. In 1854 they wrote:

> Great commiseration is, no doubt, due to many who come within this category; for we can fully imagine how shame and anguish must weigh on an unfortunate and betrayed female, with enfeebled system, what strong temptations induce her to evade the censure of the world in the destruction of the evidence of her guilt, by a crime that outrages her most powerful instinct, maternal love of offspring. The thought of such a fearful exposure no doubt may lead to some sudden and impulsive act, for which, as generally happens, she is judged with the utmost leniency.[34]

This extract also reflects one of two contradictory views of women who killed their infants. These two views are discussed thoroughly by Mark

Jackson, in his study of new-born child murder in eighteenth-century England.[35] One view saw the woman as a wicked murderer and the other saw her as a virtuous victim attempting to protect her honour. The court's job was to decide which type of woman stood before it awaiting judgement. If she was the first kind, she deserved to be severely punished, but, if the second, she deserved to be cared for. Jackson argues that as the eighteenth century drew to a close in England, the humanitarian approach gained some ground and led to debates on 'temporary insanity' linked to childbirth and the emergence of the diagnosis of 'puerperal insanity'.

In the Dundrum Asylum medical records, the diagnosis of puerperal insanity was used in relation to only one of the women found to be insane at the time she killed her infant (see Table 5.1). This was Ellen Byrne, a 26 year old Catholic prostitute from Dublin, whose case was covered in the newspapers as 'the Goldenbridge infanticide'. Ellen had drowned her baby in the canal. She said that she had been refused entry to the South Dublin Union and that she had tried to give the baby away but, when she did not succeed, she left the baby at the side of the canal, hoping someone would find it and care for it. She was found 'guilty but insane' and sent to Dundrum Asylum in 1893. In her medical notes, she was described as a 'nervous and delicate subject, greatly emaciated and run down constitutionally through her mode of life, privation and drink', and later as 'a bloodless emaciated girl, looks like a mere child, is the subject of advanced phthisis (tuberculosis)'. As well as being physically debilitated, she was 'depressed, apathetic and emotional' at the time of her admission to Dundrum Asylum. She was diagnosed as having 'puerperal insanity' and died in 1894, within a year of admission.[36] As there was no debate in the records on puerperal mania in relation to Ellen, we have to rely on the general medical literature of the time to provide us with the background to this diagnosis.

Insanity and childbirth

The debate on the impact of childbirth on a mother's mental state and the possibility of it leading to irrational behaviour, including violence, was part of a wider discourse on female insanity.[37] This debate was based on assumptions about the female psyche, which was understood to have characteristics such as passivity, emotional instability and irrationality.[38] All of these characteristics were seen as highly related to the reproductive cycle, which left women vulnerable to periods of mental disorder. The weeks before and after childbirth constituted one such period of vulnerability. Medical opinion in mid-nineteenth-century

Ireland is reflected in an article on 'mental disorders of pregnancy and childbed' by Dr Fleetwood Churchill, published in the *Dublin Quarterly Journal of Medical Science* in 1850.[39] He set the tone for his essay by elaborating on the complexity of the relationship between mind and body: 'the cords are so fine and so tense that an excess of vibration in the one extremity induces discord in the other'.[40] Having made the general point about the delicacy of the inter-relationship between mind and body, Churchill went on to elaborate his theory in relation to women: 'how much more exposed must women be to such disturbances ... the subjects of repeated constitutional changes and developments of a magnitude and importance unknown to the other sex'.[41]

There was nothing new about this view of women but Churchill used it to set the scene for his next argument, which made the connection between insanity and physical conditions related to reproduction, such as 'menstruation, conception and pregnancy, parturition and childbed, and lactation'.[42] This kind of argument formed the intellectual background for legal and medical decisions made in relation to Irish women involved in crime during the nineteenth century. The more enlightened Irish psychiatrists were active members of the Royal Medico-Psychological Association and read the *Journal of Mental Science*, which contained the writings of such well known British psychiatrists as Dr Henry Maudsley.[43] In an article on the association between menstruation, pregnancy and homicidal tendencies in 1863, Maudsley wrote:

> Irregularities of menstruation, as recognised causes of nervous disorder, may act on different parts of the nervous system in different persons, in one giving rise to hysterical convulsions or hysterical mania, in another to epilepsy, and in another to suicidal or homicidal impulse ... Morbid impulses notably spring up during pregnancy.[44]

Examples of this argument had already appeared in the very early reports on Dundrum Asylum. For example, in the annual report for 1852 we read:

> In the Asylum at Dundrum there are a few cases, which we trust may soon become subjects for your Excellency's benevolent consideration: amongst others, that of a young woman, of respectable condition, and the mother of three children, who, from fright at her

last confinement, was attacked by puerperal mania, and destroyed her infant. She is now, and has been for about eighteen months restored to reason; her husband and family are urgent for her liberation.[45]

This argument was also used in evidence to the Royal Commission on Capital Punishment in 1866, as part of the debate on the appropriateness of the death sentence for women who had killed their children.[46] The Home Office in England, in 1864, had adopted the policy of advising that the death penalty be commuted in all cases where a mother had been convicted of murdering her new-born baby (defined as being under one year old).[47] This policy was also extended to Ireland, so that women who were found guilty of the murder of a new-born child and sentenced to death, had no difficulty in having their sentences commuted to penal servitude for life. However, there was a more compassionate route open if a woman's crime could be linked to insanity. Dr F.X. MacCabe, medical officer of the General Prisons Board, in his 1886 report to the chairman of the Board on women prosecuted for new-born child murder, argued that

> If Puerperal Mania could be ascertained to have existed in any of these cases, that circumstance would totally alter their aspect from a medico-legal point of view, as it is well known that in Puerperal Mania no symptom is more constant than a perversion of the maternal instinct leading to the destruction of the infant.[48]

The debate on 'puerperal mania' as a cause of violence towards new-born babies, made it possible for courts to excuse a woman of responsibility for her actions in killing her child. Defined as criminal lunatics, there was no time-limit on the period that these women could be confined in Dundrum Asylum, whether they were judicially ruled 'unfit to plead' or 'guilty but insane'.[49] Discharge was at the pleasure of the lord lieutenant and depended on a return to sanity and a supportive family. Some of these women, like Margaret Rainey and Hannah Sullivan, were discharged within one or two years of committing their crimes. Others were destined to spend many years within an asylum, either in Dundrum Asylum or in their local district asylum. These were the women who were diagnosed as having chronic mental disorders, such as mania, melancholia, dementia and insanity related to overindulgence in alcohol.

Child murder and chronic insanity

In the annual report from the inspector of lunacy on Dundrum Asylum for 1854, the difficulties surrounding the detention of women sent to them for child murder were discussed:

> Unless the deed is accompanied with, and followed by distinct symptoms of insanity, the difficult question presents itself to us: is such a person – sane immediately after the act, sane at trial, and sane on admission – to remain for life – or, if not for life, for what period – the inmate of an asylum, and the associate of lunatics?[50]

As in the case of other patients in Dundrum Asylum, the decision of the court to excuse individuals from responsibility for a crime, on the grounds that they were suffering from temporary insanity at the time, did not mean that these people continued to be of unsound mind after conviction. Not all of the female patients who had killed children were suffering from a temporary insanity such as 'puerperal mania'. In fact, many were judged to have a chronic condition that simply manifested itself at the time of the crime, a theory expounded by Dr C. Lockhart Robertson, of the Sussex Lunatic Asylum, in the *Journal of Mental Science*.[51] His arguments are reflected in the medical notes of some of the Dundrum women who killed one or more of their children (for diagnoses, see Tables 5.1 and 5.2). These women never recovered their sanity and continued to be incarcerated in Dundrum or in another asylum until they died.

Some of them had killed a new-born child and some had killed an older child. For example, Mary Glass, a 21 year old single Catholic girl from county Antrim, was found 'unfit to plead' in 1868 in relation to the murder of her new-born child.[52] Mary was judged to be a 'congenital idiot'; in other words, she was intellectually disabled. She had the added disadvantage of her mother being a patient in Dundrum Asylum, although we do not know for what crime or condition the mother was admitted. Mary spent 34 years in Dundrum Asylum and was then transferred to Antrim district asylum in 1902. This seems an inordinately long time in confinement as a criminal lunatic for the crime of infanticide, which usually attracted a much shorter period of detention. While one might conjecture that the presence of Mary's mother in Dundrum Asylum delayed her transfer to the less stigmatising environment of the district asylum, this was not the case, as the mother had died in 1873.[53] We can only assume that Mary fell into the category of patients who were found difficult to handle, either because they were a danger to themselves or to others. Medical records tell us very little about Mary and she may have just been overlooked by the resident medical superintendent who had responsibility for transferring and discharging patients. During her time in Dundrum Asylum, Mary would have been under the care mainly of Dr Isaac Ashe, who held the position of resident medical superintendent from 1872 to 1892. Dr Ashe's methods of managing the asylum were highly questionable and finally led to an enquiry by the inspectorate of lunacy in 1891 and to his replacement in the following year by Dr George Revington.[54] During the 1890s, efforts were made to speed up discharges and transfers, but Mary was not among the lucky ones.

Ann McDonnell, another woman whose time in Dundrum Asylum was longer than average, was diagnosed as suffering from melancholia for most of her time there. Ann was a 38 year old Catholic married woman from County Sligo (see Table 5.1).[55] Ann was married at the age of 20 and had eight children. When a ninth child was born, she suffered an attack of 'acute melancholia' and killed the baby by drowning it in a nearby river. She was indicted for murder and found 'guilty but insane'. Ann spent ten years in Dundrum Asylum before being discharged to her husband in 1910. She was assessed as sane at the time of her discharge.

Some women with chronic conditions were never released from Dundrum Asylum. One of these was Mary O'Flaherty, a 34 year old Catholic married servant, who had given birth to six children. Five of her children died young from natural causes. When her sixth baby appeared to be delicate also, she drowned it. The baby was just eight months old. Her medical notes state that 'for the last seven years, she had neglected her religious duties and now, fearing the death of her remaining child, she thought she was doomed'. She made up her mind to drown herself and the baby but she survived while the child did not. She was found to be insane at the time of the crime and sent to Dundrum Asylum in 1892. She remained there until her death and, though her husband came regularly to see her, 'she took no pleasure in his visits'.[56] There is no discussion in the medical files as to why some women, such as Mary, were deemed to be insane at the time of their crimes. For example, we do not know if the fact that she had attempted suicide had any impact on the assessment. What we do know is that for those who showed little or no sign of insanity after admission to Dundrum Asylum, discharge within a very short time was a real possibility.

Two other women, who killed more than one child, were regarded by medical opinion as suffering from chronic conditions. They were Sarah McAlister, a 33 year old Catholic married woman from county Antrim, who poisoned the youngest two of her six children in 1892, and Catherine Wynn, a 35 year old Catholic married woman from county Sligo, who drowned her three children in a bath of boiling water in 1893. Sarah McAlister denied killing her children, but admitted that she tried to kill herself because she had learned of her husband's infidelity. According to her medical notes, she continued in her efforts to commit suicide: 'we have great trouble getting her to eat food ... she wants to starve herself'. Catherine Wynn had also tried to kill herself by putting her head into the same bath of boiling water that she had used to drown her children. Neither of these women recovered their sanity.[57]

We know very little about several other women who were found to be insane at the time of their crimes, because they were admitted directly to district asylums and did not appear in Dundrum Asylum records. For example, an English barrister, Michael J.F. McCarthy, writing in 1901 on 'religious insanity', referred to a Mrs Sadlier from county Tipperary, who killed her four daughters aged between five months and four and a half years in 1896. She cut their throats with a razor.

> Their heads were almost severed from their bodies .... Her words were, as sworn to by the Sergeant at the inquest: 'Well, I killed the four children in order that they may be with the Almighty God, as I consider they were not capable of committing sin. I hope they were not. They were not up to the age of reason. I strove to destroy them before they would fall into the same sins that I had committed'.[58]

McCarthy selected this case to highlight the evils of religious fervour, but his account is verified in Royal Irish Constabulary (RIC) records, which tell us that Mrs Sadlier appeared to be 'suffering from religious mania', though she had never shown any signs of insanity before she killed her children. She was sentenced and remanded and was later 'removed to Limerick district asylum by order of the lord lieutenant'.[59] Like Mary O'Flaherty, Sarah McAlister and Catherine Wynn, Mrs Sadlier never recovered mentally from this tragedy.

What is interesting for our discussion here is that these women were deemed 'mad' rather than 'bad' and were thus excused of responsibility for this most terrible crime. In her study of folklore surrounding child murder, Anne O'Connor suggests that for women who killed more than one child, repentance was the key to being accepted by society. If a woman knew that she had done wrong and was sorry for it she could be forgiven. Otherwise she left herself open to being completely ostracised as evil:

> In our examination of child murderess traditions, the figure of a Satanic woman who would destroy her own child/children or that/those of another woman without repenting of such an act, has emerged. This figure has been seen to resemble the witch-midwife character in her diabolical association and evil deeds.[60]

Accounts of women sent to Dundrum Asylum reveal that it was not quite so simple. Many women did not repent of their crimes, as they did not appear to understand the gravity of their actions. However, they were not treated as evil. In fact, many were excused of responsibility on the grounds of insanity and were cared for rather than punished. Care in an asylum, however, while better than confinement in a prison, was still characterised by loss of liberty and rejection by society.

A murder conviction

Before concluding this chapter, it is important to remember that the women for whom the insanity defence was used did not represent the whole population of women who came before the courts for killing an infant or young child.[61] For a successful insanity plea, there had to be some indication that this was a genuine case of insanity rather than a deliberate and calculated act of destruction. As shown above, in the report of the inspector of lunacy on Dundrum Asylum for 1854, the difficulties surrounding the conviction and the sentencing of these women were discussed.[62] These difficulties often led to the early release of the women in Dundrum Asylum who had committed infanticide. However, as convict records of the period show, many other women were found guilty of murder or manslaughter and sentenced to execution or penal servitude. Luckily for them, there was no tradition of executing women for child killing during the second half of the nineteenth century. Neither were they all compelled to stay in prison for life. In fact, convict records indicate that many were freed long before their prison term expired. The length of time spent in prison varied between three and 15 years, a matter that was the subject of debate later in the century.[63] Afterwards, some went home to their families and some emigrated directly from prison to the US or Canada. Like other ex-convicts during this period, if they so wished, they were delivered directly to the docks (usually Liverpool), having received a sailing ticket for a specific boat and the money they had earned during their time in prison.[64]

As the nineteenth century drew to a close, it was becoming increasingly clear that there was great opposition to the death penalty for women, especially when this sentence was imposed for the killing of a young child. In spite of this opposition, the law laid down that a deliberate act of killing another person, even if it was a baby, incurred the sentence of death. Only after conviction was the sentence usually reduced. In 1886, there was intensive correspondence on the topic between government departments at Dublin Castle. Following a directive from the Home Office in London, the lord lieutenant requested reports on all the women who were in prison at the time for the killing of their children. This provided an opportunity for those officials who considered the death sentence unjust to place their arguments in writing before the highest authority in the land. The two most powerful voices in this debate were the Prison Board's Chairman, Mr F. Bourke, and Medical Officer, Dr F.T. MacCabe. They reported that between 1864 and 1883, 14 women had been found guilty of murdering their children and sentenced to be hanged and they raised two major concerns about the legal treatment of these women, none of whom had been executed. The first issue was that of the great variation in the final sentences handed down to these women. Most were initially sentenced to death, but when this was commuted, some were sentenced to as little as 12 months penal servitude, while others were sentenced to life-long imprisonment 'without any appearance of hope ... for crimes that appear to be far less heinous'.[65] The second reason put forward by Dr MacCabe was that his research indicated that Ireland seemed out of step with other countries in Europe in its punitive approach to infanticide. Not only were these women sentenced to death and later condemned to imprisonment, but they were also rejected by their families and communities. Despite these concerns, expressed in the 1880s, the law in the south of Ireland, was not changed until over 60 years later, with the passing of the Infanticide Act in 1949.[66]

Conclusion

We have looked at one of the responses in nineteenth-century Ireland to the killing of an infant or young child by a mother: namely, the insanity defence, which diverted these women from execution and prison to the mental health system of the time. During the 1840s and 1850s, the mental health system in Ireland expanded and improved. An inspectorate of lunacy was formed in 1845 to oversee existing district asylums and to plan for future expansion. Dr Francis White, who had held the position of inspector of lunacy within the prison system, was put in charge of the new inspectorate. One of the most important developments in those early years was the planning and building of a criminal lunatic asylum at Dundrum, county Dublin, which opened in 1850. Psychiatrists, from the inspectorate of lunacy and from Dundrum Asylum, played a major role in determining who was accepted into this very limited facility of 100 to 120 beds. Women were a minority in Dundrum Asylum, but a substantial number of its female patients had been involved in the killing of their infants or young children. In determining which women needed this type of care, Irish psychiatrists relied on Britain for most of their ideas and training. Medical records from the second half of the nineteenth century are not very detailed, probably as a result of the very low doctor/patient ratio in the asylum; medical staff at Dundrum usually consisted of two resident psychiatrists (the resident medical superintendent and his assistant) and one visiting physician. Fortunately, we are able to supplement these medical records with a fairly detailed annual report which was published by the inspectorate of lunacy each year. Some of this was written by the inspector who had carried out a visit, and some by the resident medical superintendent. From these reports, we can learn something of the ideas behind the decisions that were made. There are also some medical reports in official files held by Dublin Castle on patients who appealed for early discharge. From this mixture of material, we conclude that Irish psychiatry had a distinctive and important role in the criminal justice system in the second half of the nineteenth century.

However, psychiatrists worked within the context of wider developments in medical practice in nineteenth-century Ireland. Within medicine, psychiatry did not have high status, but in legal circles its place and status were firmly established by the opening of the Central Criminal Lunatic Asylum at Dundrum. With an average of 100 beds available for criminals whose insanity made them difficult to cope with in the ordinary prison system, Dundrum Asylum provided an escape route for men and women who might otherwise have faced the death penalty. It also provided opportunities for Irish doctors interested in studying the impact of mental disorder on criminal behaviour.[67] Together with their colleagues from the district asylums, the doctors at Dundrum Asylum developed the theoretical and administrative bases for the future of Irish psychiatry.

Notes

1 K. Daly, *Gender, Crime and Punishment* (New Haven, 1994); A. Morris, *Women, Crime and Criminal Justice* (Oxford, 1987). For a discussion on the emergence of psychiatry as a specialism in the nineteenth century see John Gach, 'Biological psychiatry in the nineteenth and twentieth centuries', in Edwin Wallace and John Gach (eds), *History of Psychiatry and Medical Psychology* (New York, 2008), pp. 381–418.
2 Inez Bailey, 'Women and crime in nineteenth-century Ireland' (MA thesis: St Patrick's College, Maynooth, 1992); Robert Bluglass, 'Infanticide and filicide', in R. Bluglass and P. Bowden (eds), *Principles and Practice of Forensic Psychiatry* (Edinburgh, 1990); S. Jackson, 'Gender, crime and punishment in late nineteenth-century Ireland: Mayo and Galway examined' (MA thesis: National University of Ireland, Galway, 1999).
3 For a discussion on terminology, see Ian O'Donnell, 'Lethal violence in Ireland, 1841 to 2003', *British Journal of Criminology*, 45 (2005), pp. 671–95.
4 Infanticide Act 1922 (England and Wales) 12 & 13 Geo. 5 c. 18; Infanticide Act 1938 (England and Wales) 1 & 2 Geo. 6 c. 36; Infanticide Act 1949 (Ireland) 1949, No. 16.
5 They were sentenced according to the current legislation on criminal lunacy: Criminal Lunatics Act 1800, 39 & 40 Geo. 1 c. 12; Criminal Lunatics (Ireland) Act 1838, 1 & 2 Vic. c. 27; Central Criminal Lunatic Asylum (Ireland) Act 1845, 8 & 9 Vic. C. 107; for discussion, see Pauline M. Prior, *Madness and Murder: Gender, Crime and Mental Disorder in Nineteenth-Century Ireland* (Dublin, 2008), pp. 50–80.
6 Minister's papers, Dail, Infanticide Bill 1949. JUS/90/8/218, National Archives of Ireland (hereafter NAI).
7 I would like to acknowledge the financial support for this research project from the Wellcome Trust, London, and to thank Dr Art O'Connor (formerly of Central Mental Hospital, Dundrum), Gregory O'Connor (NAI), and Norma Menabney of Queen's University Library.
8 Roy F. Foster, *Modern Ireland 1600–1972* (London, 1988); Leslie A. Clarkson and E. Margaret Crawford, *Feast and Famine: A History of Food and Nutrition in Ireland 1500–1920* (Oxford, 2001); William E. Vaughan, 'Ireland c.1870', in W.E. Vaughan (ed.), *A New History of Ireland: V, Ireland under the Union, 1, 1801–70* (Oxford, 1989), pp. 726–802.
9 Letter from Dr MacCabe to the Chairman of the Prisons Board, 30 June 1886, CRF Misc., 1888/no. 1862, NAI.
10 Sean Connolly, 'Illegitimacy and pre-nuptial pregnancy in Ireland before 1864: The evidence of some Catholic parish registers', *Irish Economic and Social History*, 6 (1979), pp. 5–23; O'Donnell, 'Lethal violence'; R. Sauer, 'Infanticide and abortion in nineteenth-century Britain', *Population Studies*, 32:1 (1978), pp. 81–93.

11 Dympna McLoughlin, 'Infanticide in nineteenth-century Ireland', in Angela Bourke et al. (eds), *Field Day Anthology of Irish Writing, Vol. 4: Irish Women's Writing and Traditions* (Cork, 2002), pp. 915–22.
12 O'Donnell, 'Lethal violence', Fig. 1, p. 677.
13 Bluglass, 'Infanticide and filicide', p. 524.
14 Ibid.
15 Liam Kennedy, 'Bastardy and the great famine: Ireland 1845–1850', *Continuity and Change*, 14:3 (1999), pp. 429–52; O'Donnell, 'Lethal violence'.
16 For discussion of legal basis for admissions to Dundrum Asylum, see P. Gibbons, N. Mulryan, A. McAleer and A. O'Connor, 'Criminal responsibility and mental illness in Ireland 1850–1995: Fitness to plead', *Irish Journal of Psychological Medicine*, 16:2 (1999), pp. 51–6; P. Gibbons, N. Mulryan and A. O'Connor, 'Guilty but insane: The insanity defence in Ireland 1850–1995', *British Journal of Psychiatry*, 170 (1997), pp. 447, 467–72; Pauline M. Prior, 'Crime, mental disorder and gender in nineteenth-century Ireland', in I. O'Donnell and F. McAuley (eds), *Criminal Justice History: Themes and Controversies from Pre-Independence Ireland* (Dublin, 2003), pp. 66–82.
17 Correspondence between Prisons Board and Dublin Castle during the 1880s, CRF, Misc., 1888/no. 1862, NAI.
18 Letter of 1 July 1886 from the Chairman of the Prisons Board to the Under Secretary, Dublin Castle, CRF, Misc., 1888/no. 1862, NAI; Report to the CSO on females sentenced to death since 1881, CRF, 1902/ D76 (Daly), NAI.
19 Honora and Bridget Stackpoole in 1853, Margaret Sheil in 1870, GPB CN5: Death Book 1852–1930, NAI.
20 Roger Smith, *Trial by Medicine: Insanity and Responsibility in Victorian Trials* (Edinburgh, 1981), p. 147.
21 Ellen Ross, *Love and Toil: Motherhood in Outcast London 1870–1918* (New York and Oxford, 1993), p. 187.
22 Niamh Mulryan, Pat Gibbons, Art O'Connor, 'Infanticide and child murder – Admissions to the Central Mental Hospital 1850–2000', *Irish Journal of Psychological Medicine*, 19:1 (2002), pp. 8–12, Fig. 1, p. 10.
23 Mark Finnane, *Insanity and the Insane in Post-Famine Ireland* (London, 1981); Arthur Williamson, 'The beginnings of state care for the mentally ill in Ireland', *Economic and Social Review*, 1:2 (January 1970), pp. 281–90.
24 *Report of the Commissioners of Inquiry into the State of Lunatic Asylums ... in Ireland*, H.C. 1857–58 (2436), xxvii, 1; W.E. Vaughan and A.J. Fitzpatrick, *Irish Historical Statistics: Population, 1821–1971* (Dublin, 1978), p. 3.
25 Carolyn Conley, *Melancholy Accidents* (Lanham Md., 1999); Elizabeth Malcolm, 'Investigating the "machinery of murder": Irish detectives and agrarian outrages, 1847–70', *New Hibernia Review*, 6:3 (2002), pp. 73–91; O'Donnell, 'Lethal violence', p. 691; Vaughan, *A New History of Ireland V*, p. 764.
26 Robert Menzies, 'Contesting criminal lunacy: Narratives of law and madness in west coast Canada, 1874–1950', *History of Psychiatry*, 12 (2001), pp. 123–56; Ralph Partridge, *Broadmoor: A History of Criminal Lunacy and its Problems* (London, 1953).
27 *Report of the Inspectors of Lunacy on the District, Criminal and Private Asylums in Ireland*, H.C. 1847 (820), xvii, 355, 362 (Hereafter, *Asylums Report*);

Pauline M. Prior, 'Prisoner or patient? The official debate on the criminal lunatic in nineteenth-century Ireland', *History of Psychiatry*, 15 (2004), pp. 177–92.

28 *Asylums Report*, p. 146, HC 1866 (3721), xxxii, 125.
29 CRF, 1894/ R.10 (Rainey), NAI; Dundrum Asylum, Female Casebook, M. Rainey, 1891, no. F838, p. 65; Dundrum Asylum, Physician's Book, p. 226.
30 Dundrum, Female Casebook, H. Sullivan, 1895, no. F946, p. 145.
31 Margaret Arnot, 'Understanding women committing new-born child murder in Victorian England', in Shani D'Cruze (ed.), *Everyday Violence in Britain 1850–1950* (London, 2000), pp. 55, 60.
32 Cited in ibid., p. 60.
33 Dundrum, Female Casebook, H. Sullivan, 1895, no. F946, p. 145.
34 *Asylums Report*, p. 155, HC 1854–55 (1981), xvi, 137.
35 Jackson, *New-Born Child Murder*, pp. 111–28.
36 Dundrum, Female Casebook, E. Byrne, 1893, no. F874, p. 93; Gibbons et al., 'Guilty but insane', p. 469.
37 Elaine Showalter, *The Female Malady: Women, Madness and English Culture 1830–1980* (London, 1987).
38 Ibid.; Smith, *Trial by Medicine*, p. 143.
39 For further discussion on Dr Fleetwood Churchill and on puerperal insanity, see Hilary Marland, *Dangerous Motherhood: Insanity and Childbirth in Victorian Britain* (Basingstoke, 2004).
40 Fleetwood Churchill, 'On the mental disorders of pregnancy and childbed', *Dublin Quarterly Journal of Medical Science*, xvii (February 1850), pp. 38–63, p. 39.
41 Ibid.
42 Ibid.
43 Mark Finnane, 'Irish psychiatry, part 1: The formation of a profession', in G.E. Berrios and H. Freeman (eds), *150 Years of British Psychiatry 1841–1991* (London, 1991), pp. 306–13; David Healy, 'Irish psychiatry, part 2: Use of the Medico-Psychological Association by its Irish members – plus ca change!', in Berrios and Freeman, *150 Years*, pp. 314–20.
44 Henry Maudsley, 'Homicidal insanity', *Journal of Mental Science*, ix:47 (October 1863), pp. 327–43.
45 *Asylums Report*, HC 1852–53 (1653) xli, 353, 368.
46 *Report of the Royal Commission on Capital Punishment 1866, together with the Minutes of Evidence and Appendix*, HC 1866 (3590) xxi, 1.
47 Ross, *Love and Toil*, p. 187.
48 Letter from Dr MacCabe to the Chairman of the Prisons Board, 30 June 1886, CRF, Misc. 1888/no. 1862, NAI.
49 See note 3.
50 *Asylums Report*, HC 1854–55 (1981) xvi, 137, 155.
51 C. Lockhart Robertson, 'A case of homicidal mania, without disorder of the intellect', *Journal of Mental Science*, vi:34 (July 1860), p. 395.
52 Dundrum, Female Casebook, M. Glass, 1868, no. F351, p. 5; Gibbons et al., 'Criminal responsibility', p. 53.
53 Gibbons et al., 'Criminal responsibility', p. 53.
54 For discussion on Dr Ashe, see Prior, *Madness and Murder*, pp. 69–75.
55 Dundrum, Female Casebook, A. McDonnell, 1900, no. F1052, p. 189.

56 Dundrum, Female Casebook, M. O'Flaherty, 1892, no. F853, p. 77. Place of origin and date of death unknown.
57 Dundrum, Female Casebook, S. McAlister, 1892, no. F861, p. 81; C. Wynn, 1893, no. F868, p. 85.
58 M.J.F. Mc Carthy, *Five Years in Ireland 1895–1900* (Dublin, 1901), pp. 190–1.
59 *RIC Return of Outrages for 1896*, Homicides, p. 11, Police Reports 1882–1921, Box 4, NAI.
60 A. O'Connor, *Child Murderess and Dead Child Traditions: A Comparative Study* (Helsinki, 1991), p. 103.
61 For further discussion, see Prior, *Madness and Murder*, pp. 131–42.
62 *Asylums Report*, H.C. 1854–55 (1981) xvi, 137, p. 155.
63 CRF, Misc. 1888/no. 1862, NAI.
64 For example, see PEN, 1886/105 (Aylward) and PEN, 1886/239 (Pritchard), NAI; for further discussion, see Prior, *Madness and Murder*, pp. 207–24.
65 Letter of 1 July 1886 from the Chairman of the Prisons Board to the Under Secretary, Dublin Castle, CRF, Misc. 1888/no. 1862, NAI.
66 Minister's papers, Dail, Infanticide Bill 1949, JUS/90/8/218, NAI.
67 Prior, 'Prisoner or patient?', pp. 177–92.

6
Science, Politics and the Irish Literary Revival: Reassessing 'Dr Sigerson' as Polymath and Public Intellectual

James McGeachie

Introduction

George Sigerson (1836–1925) is now an obscure and largely forgotten figure.[1] He left no memoir, and his personal papers and manuscripts are apparently lost. To date, there has been no biography to give an appropriately detailed account of his long life and career. Neither is there a monograph providing an extended study of his work as a medical man, a man of science and a public intellectual. His name is best known today through the Gaelic Athletic Association (GAA), either in the form of the intercollegiate Sigerson Cup competition in Gaelic football he endowed in 1911, the Strabane Sigersons club of his hometown, or as the author of the Tyrone anthem 'The Mountains of Pomeroy'.[2] In his day, however, the Strabane-born Sigerson enjoyed a high public profile, both as a *savant* and as a poet, historian and political journalist.

An eminent figure in the bio-medical sciences in Ireland, Sigerson combined a fashionable Dublin practice with a long and distinguished academic career researching and teaching botany and zoology at the Catholic University Medical School and its successor institutions, the Royal University of Ireland and University College Dublin (UCD). As a clinician, his particular bent was towards the emergent medical specialism of neurology. He studied in Paris under Jean Martin Charcot, the founder of modern neurology and Freud's mentor in the study and treatment of hysteria through the talking cure. As favoured students of Charcot, Sigerson and Freud were invited to translate selections of Charcot's *Leçons* into English and German respectively. Sigerson also published articles on a variety of scientific and medical topics in the leading Irish and British journals in these fields. As a prominent Dublin naturalist, he also took part in the Irish debate about Charles Darwin's theory of evolution by natural selection, corresponded with Darwin and was nominated by the latter for membership of the élite Linnaean Society of London.

Sigerson was also one of the leading figures of the Irish Literary Revival (hereafter the 'Revival') of the late nineteenth and early twentieth centuries to which he contributed as a poet, an anthologist, a translator of Irish language poetry into English, a defender of the Irish language, and a champion of the openness and inclusiveness of Irish culture. Between the 1860s and the 1900s he was politically active in relation to much that was going on in Irish nationalism. Initially he did so as a journalist in the radical nationalist coteries around *The Shamrock* and other journals where the residual remnants of Young Ireland rubbed shoulders with Fenians. Later he became a respected public commentator on Fenianism, the land question and Irish political prisoners. The views of this later Sigerson were listened to with respect by the English political and intellectual élite. Through the poet and novelist Hester Varian, Sigerson married into the political and literary legacy of Young Ireland. Their daughter, Dora Sigerson Shorter maintained the familial link between radical nationalism and artistic production. A sculptor, artist and poet, Dora worked on the 1916 memorial in the chapel of Glasnevin Cemetery and wrote *The Tricolour* (1922), a collection of poems eulogising the leaders of the Easter Rising.[3]

Within the social history of medicine in Ireland, the life of George Sigerson is notable for the ways in which it illustrates how a particular type of career and public role for a medical man was possible within the culture of care of the Dublin medical world in the later nineteenth and early twentieth centuries. In that culture of care, Sigerson's self-fashioning could still be formulated in the terms of a polymathic model, which predated the professionalisation and the scientism that were coming to the fore in Irish medicine, as elsewhere, at that time. As well as embodying some of the distinguishing features of intellectual life in the Ireland of his time, Sigerson's role as a man of science and a public intellectual can also be seen as contributing part of an Irish dimension to the historiography of national styles of science. Combining a career in the bio-medical sciences with a lifetime of interventions as a public intellectual, Sigerson was part of an Irish medical lineage that also includes Sylvester O'Halloran (1728–1807), Richard Robert Madden (1798–1886) and Sir William Wilde (1815–1876).

A new assessment of Sigerson, integrating his role in the Revival and in contemporary nationalist polemic with his career in science and medicine, is well overdue. This essay will make some suggestions as to how such a reassessment might be made. It will begin with an outline of the disparate but frequently interconnected strands of the career trajectory of this latterly neglected late-Victorian phenomenon, invariably referred to by contemporaries as 'Dr. Sigerson'. Sigerson's public reputation, and the high esteem in which he was held in his final years, will then be contrasted with the more recent scholarly neglect of him. The significant absences of Sigerson from recent studies of the cultural history of the Revival will be noted. This will be followed by an examination of how Sigerson has been represented in two discrete areas of scholarship: in the essays on the history of Irish medicine by the late J.B. Lyons, and in the interdisciplinary work in Irish studies by the Field Day cohort. With reference to these two bodies of work, it will be suggested that Sigerson's particular combination of a career in the bio-medical sciences with a prominent role in the Revival and in the wider nationalist polemic of the period add significant dimensions to our understandings of Irish medical history, of the Revival, and of the culture of intellectual life in modern Ireland.

George Sigerson was the youngest son of a substantial Strabane family whose prosperity came both from inherited land and from manufacturing.[4] From his parents' religiously mixed marriage, Sigerson was also linked by blood with two of the primary historic strands of Irish national memory.[5] Through his mother, a protestant Nielson of Strabane who was related to the United Irishman Samuel Nielson, there was a living connection both with the non-sectarian republican project of the United Irishmen and with the regional particularism of the politics of the north-west of Ireland.[6] And with a Derry Catholic father whose family had moved there from Kerry, there was also a family link with the displaced native Irish of seventeenth-century Munster and, further back, with the old Pale families of medieval and early modern Dublin. Douglas Hyde notes that 'one Christopher Sigerson appears amongst the list of the transplanted Irish in 1654 and Dr Sigerson was probably of the same family ... If so, heredity would sufficiently account for his strong national sympathies'.[7]

Those 'strong national sympathies', however, were supplemented from an early stage of Sigerson's schooldays by the beginnings of a lifelong engagement with France. After initial schooling at the Glebe school in Strabane and at Letterkenny Academy under the latter's Franco-Hibernian headmaster, a Dr Grenand, Sigerson moved to France itself to attend the Collège St Joseph, in what was then the new Parisian suburb of Auteuil. Here he completed his education during Napoleon III's first years as French Emperor, and won the school prize for his Latin verse translation of 'The exile of Erin', the award being presented to him by the Grand Aumonier of the Emperor. After Auteuil, Sigerson returned to Ireland in 1855 as a student, initially at Queen's College, Galway, but transferring the following year to its sister college in Cork, with a scholarship to study medicine.[8] Alongside medicine, Sigerson also studied the Irish language in Cork, taking First Honours in Irish at the same time as his M.D. He mixed in nationalist circles there, notably with the residual Young Ireland cohort centred around the poets Ralph and Isaac Varian,[9] to whose sister Hester, herself a poet and novelist, Sigerson became engaged. Sigerson's 'The mountains of Pomeroy' was one of the ten poems he later contributed to Ralph Varian's 1869 collection *The Harp of Erin*.[10] It was during his Cork period that Sigerson started writing for *The Nation*. In 1859, the year of his Queen's University of Ireland MD and the year in which Charles Darwin was panicked into finally publishing his conclusions on the workings of biological evolution by natural selection as the *Origin of Species* was also the year in which Sigerson published his first book, under the pseudonym 'Erionnach, MD'. This was a translation, with preface, of 46 poems into English from Irish for the second series of John O'Daly's *The Poets and Poetry of Munster*. The first series had been done ten years earlier by James Clarence Mangan, who had died by this time.

Sigerson's final year of preparation for his MD had been spent studying surgery at the Catholic University School of Medicine in Cecilia Street, Dublin, and after marrying Hester Varian in 1861, Sigerson set himself up in a Dublin medical practice, initially at 17 Richmond Hill, Rathmines. In 1865, after being awarded an MCh from the Queen's University of Ireland,[11] he joined the expanding professorial staff of the Catholic University School of Medicine. This began a close association with the School which would last through its various institutional transformations until 1923. As William Doolin observed, Sigerson served the Catholic University 'during the lifetime of the Royal (University) and University College Dublin ... Over the years the title of his appointment changed, but he was essentially Professor of Botany and Zoology.'[12]

It was in the latter capacity that, from the early 1860s to the mid-1880s, Sigerson contributed papers to the science series of the *Proceedings of the Royal Irish Academy* on a variety of scientific topics. Some of these dealt with matters of specifically Irish interest, like the flora of one of the botanical districts into which Ireland was divided and 'Fish-remains in the alluvial clay of the River Foyle'.[13] Others, however, addressed some of the central concerns of later nineteenth-century science and of its leading public champions. With 'On heat as a factor in vital action' (1875–77),[14] Sigerson joined the contemporary debate on the nature of heat led by the Carlow-born John Tyndall, the most publically prominent physicist of the age, and founder-member, with T.H. Huxley and Herbert Spencer, of the elite X Club of Victorian proponents of scientific hegemony, as well as being the author of the most aggressive polemic that supported the ascendancy of science in his 1874 Belfast address to the British Association for the History of Science.[15] Sigerson's 1863 article on plant physiology, 'Some remarks on a proto-morphic phyllotype', published in the Catholic University's short-lived journal of science and literature, *The Atlantis*, resulted in a cordial exchange of correspondence with Charles Darwin to whom Sigerson vouchsafed that he had 'read with so much pleasure your views on the *Origin of Species*'.[16]

It was possibly as a result of this exchange that Darwin nominated Sigerson for membership of the Linnean Society of London.[17] The correspondence itself, it should be noted, took place in the immediate aftermath of the 1859 publication of the *Origin of Species* and the very public and highly controversial clash between T.H. Huxley and Bishop Samuel Wilberforce over Darwin's book at the 1860 meeting of the British Association for the Advancement of Science in Oxford.[18] In the 1884 article, 'Considerations of the structural and acquisitional elements in dextral pre-eminence, with conclusions as to the ambidexterity of primeval man', meanwhile, Sigerson was engaging both with a specific biological debate about human and animal handedness taking place at that time and to the wider debate about human evolution and the origins of man.[19] And in his 1894 essay, 'Genesis and evolution' for an early issue of the UCD journal the *New Ireland Review*, he made a notable contribution to the wider debate in Ireland regarding Charles Darwin's theory of the evolution of species by natural selection.[20]

Sigerson's association with Darwin was indeed something that the Jesuits, who were at that time controlling UCD and with it the residual remnant of the former Catholic University and its medical school, seemed rather proud about.[21] When, in the course of testifying before the Royal Commission on University Education in Ireland in September 1901, the College President, William Delaney S.J., was asked what the position of UCD as a Catholic institution would be as regards to the teaching of Darwinian evolution, Delaney replied with reference to Sigerson's work. Delaney declared there was a 'distinguished Professor of Biology ... a great friend of Darwin – Dr Sigerson' who had sent 'to the Dublin papers an indignant protest contradicting a statement made by Mr Arnold-Forster, in the House of Commons, that a certain Professor of mine had been dismissed on account of his teaching of Biology'.[22]

Alongside the development of his medical and academic careers in the early 1860s, Sigerson's move to the capital was also marked by an intensification of his political activities. As in Cork, he was closely connected to some of the surviving remnants of Young Ireland, including The O'Donoghue and the constitutional nationalism of the short-lived Independent Party.[23] In the company of The O'Donoghue and the editor of *The Nation*, A.M. Sullivan, Sigerson travelled to Paris in 1860 with an Irish sword of honour produced in Dublin to be presented to General Patrice de MacMahon. Of Wild Geese descent, MacMahon had been created Marshall of France by Napoleon III after his 1859 victory over Austria at the battle of Magenta. In the 1870s MacMahon was to become President of the Third French Republic. At the banquet held in honour of the Marshall, which Sigerson and Sullivan attended, other descendants of Wild Geese families were also present, including Chevalier Leonard, General Sutton de Clonard, and Commandant Dillon, as well as John Mitchel, who was recently escaped from Tasmania.[24]

This visit to Paris may have been of some significance for Sigerson in two respects. In the first instance, meeting with John Mitchel at the MacMahon banquet may have influenced the markedly more radical orientation that Sigerson's nationalism subsequently took. This shift to a more advanced nationalist position than those of The O'Donoghue and Sullivan was reflected in the journalistic interventions Sigerson was making in *The Irishman* and other journals associated with Fenianism by the middle of the 1860s. Second, as J.B. Lyons has surmised,[25] being in Paris just after taking his MD may have provided Sigerson with an initial opportunity to observe at first hand the hospitals and clinics of Paris. Despite the rival claims to clinical pre-eminence of Vienna and Berlin, the French capital in 1860 arguably remained the global centre of modern, scientific medicine.[26]

It is certainly the case that while Sigerson's bent towards the emergent specialism of neurology predated the MacMahon sword episode,[27] only in its aftermath did he begin to extend his medical education with annual visits to Paris. From that time onwards, these visits enabled him to study under, and become known to, some of the leading figures in

French neurology and psychiatry. For the histology of the nervous system and for nerve anatomy, for instance, Sigerson went to Louis-Antoine Ranvier (1835–1922),[28] whose own work within the French physiological tradition drew on the methodological insights of German histology.[29] The curmudgeonly Benjamin Ball (1825–1893), expert on rheumatology at the Sainte-Anne asylum, first holder of the chair of psychiatry (*chaire de pathologie mentale et des maladies de l'encéphale*) in the École de Médecine, and author of a monograph on persecution complex,[30] became a friend of Sigerson's and may have facilitated the latter's initial meeting with Jean Martin Charcot, Ball's co-author of a volume of lectures on chronic illnesses of old age.[31] As a resident of Auteuil, however, where Sigerson had gone to school at the Collège St Joseph, Charcot may have already been known to him.

Charcot[32] and Duchenne de Boulogne[33] were to become Sigerson's closest associates in Paris. Initially, Duchenne, whose ward-based researches and case-studies of neuromuscular anatomy were of foundational importance for the understanding of hereditary muscular dystrophy, was Sigerson's mentor.[34] But above all, it was Charcot who became both a formative influence on Sigerson as a clinician and as a close friend.[35] Charcot's Tuesday clinics, accompanied by lectures at the Salpêtrière hospital for women where he was Professor of Anatomy and Pathology and, from 1882, of Neuropathology, attracted specialists from around the world and he dominated French neurology during the 1870s and 1880s. Working in the French medical tradition that gave primacy to pathological anatomy, Charcot had identified the locations of the anatomical lesions that caused diseases such as multiple sclerosis.

It is unclear when Sigerson's initial engagement with Charcot and his work occurred but Charcot's initial appointment at the Salpêtrière was in 1860, the year that Sigerson was in Paris with the MacMahon sword. According to J.B. Lyons, Sigerson was sitting in on the pathology classes that Charcot had started teaching in 1873.[36] A series of research papers on neurology, in journals published in Paris, Dublin and London, that Sigerson produced between 1874 and 1878 demonstrated his affiliation with Charcot and Duchenne. The first of these articles, 'Note sur la paralysie vasomotrice', appeared in 1874 in the *de facto* house-journal of the Salpêtrière, *Le Progrès Médical*.[37] His 1877 paper to the Dublin College of Physicians, 'On alternate paralyses', was published in the *Dublin Journal of Medical Science*[38] the following year. The 'Napoleon of the neuroses', as Charcot was sometimes known, awarded Sigerson the singular honour he gave to a select number of his Tuesday afternoon students, of inviting him to undertake an English translation of his

Leçons sur les maladies du système nerveux.[39] As *Diseases of the Nervous System*, Sigerson's translation was published in two volumes in 1877 and 1881, under the élite medical imprint of the London-based New Sydenham Society.[40] It was preceded by the *British Medical Journal's* publication of a Charcot *Leçon* in Sigerson's translation as 'Certain phenomena of hysteria major'.[41] Sigmund Freud, a pupil at the Tuesday clinics in 1885 and 1886, was similarly honoured in being invited by Charcot to undertake the German translation of the *Leçons du mardi a la Salpêtrière* (1885-6).[42]

Douglas Hyde described 'the great field of philanthropy and politics' as lying between Sigerson's dual allegiances to 'Science and Pure Literature'. In matters of science and medicine, Hyde attributed to Sigerson 'a French outlook' and 'a French attitude' respectively:

> In other words he blended theory with actual practice and applied theory to actual practice in a manner very different from that of the English or Germans who tend to dissociate these things.[43]

His medical mentors were certainly all French.[44] In addition to Charcot and Duchenne they included the zoologist Henri Milne-Edwards and the physiologist Claude Bernard whose revolutionary methodologies inspired Emile Zola's naturalistic novels. Another physiologist, Paul Bert proposed Sigerson for membership of the Clinical Society of Paris and Charcot nominated him for the Society for Physiology and Psychology. Sigerson's Parisian reputation as a man of science is demonstrated by his being invited in 1876 by the Société de Biologie to join a team of French investigators of proposed treatments for hysterical epilepsy through applications to the skin.[45] Sigerson later published accounts of these researches in the *British Medical Journal*.[46]

Through these medical mentors, Sigerson also acquired links with the human sciences and the radical intellectual coteries of Second Empire and Third Republic Paris. In particular, and through the patronage of the historian Henri Martin, he became a member of the Anthropological Society of Paris when that society was dominated by Paul Broca.[47] This association arguably puts the scepticism about Darwinism expressed by Sigerson in his 1894 article, 'Genesis and evolution', into a broader context than that in which the article was presented at the time of its publication. Appearing where the latter did, in the first issue of the *New Ireland Review*, a UCD journal produced under Jesuit patronage and represented in the editorial preface as 'an excellent specimen of the Catholic scientists' controversy',[48] the article can be seen as an antecedent of the 1902 attack on Darwin's *The Descent of Man,* by Sigerson's colleague, the physio-logist and later President of UCD, Denis Coffey,[49] and as part of a wider late nineteenth-century Catholic critique of Darwin.[50] Given Sigerson's connection with Broca and the Parisian anthropologists, however, the article may have owed as much to the relative indifference to Darwinian, as opposed to Lamarckian *transformisme,* characteristic of most French scientists and intellectuals of the period, as it did to any specifically Catholic agenda.[51]

Sigerson's Parisian connections and reputation underscored his scientific and medical integrity on a European level. Nevertheless, despite this and the eminence he gained as Charcot's translator, Sigerson never obtained a Dublin hospital appointment. This is the more remarkable given the close connections between the Catholic University School of Medicine, St Vincent's, Jervis Street, and the Mater, since no sectarian factors could have operated against the Catholic Sigerson in these institu-tions. Suspicions aroused by his nationalist politics, however, may have been a factor.[52] A hospital appointment was generally a prerequisite for admission to the medical elite. This was certainly the case in London earlier in the century as the professional frustrations suffered by the English neurologist Marshall Hall illustrate.[53] It may have been the case that medical Dublin between the 1860s and the 1880s was as unsym-pathetic towards medical specialisms as London had been in the case of Hall, although the success of Sir William Wilde's career as a specialist in diseases of the eye and ear in Dublin between the 1840s and the early 1870s suggests otherwise.[54]

The lack of a hospital appointment notwithstanding, however, Sige-rson's 1877 relocation to 3 Clare Street, close to the centre of medical Dublin in Merrion Square, where William Stokes and Sirs Dominic Cor-rigan and William Wilde, the medical luminaries of mid-Victorian Dublin, had resided and practised, does suggest that if only by proximity Sigerson had become part of the medical establishment.[55] Medical practice was indeed his main source of income, in effect bankrolling his multifarious other activities.[56] Teaching appointments at the Catholic University in the decades immediately following its establishment in 1854 were notor-iously badly paid. The path to a properly endowed scientific career, like that to a hospital appointment, may also have been obstructed by his political and literary associations, but it must have been precisely those associations that fed and made fashionable his growing practice. 'Dr Sigerson' seems to have quickly become the practitioner of choice for *haute* nationalist Dublin. His patients included Charles Kickham, at whose deathbed Sigerson was in attendance, Maud Gonne, over whose treatment he and W.B. Yeats disagreed vigorously, and John Dillon.[57] 'Old Sigerson', as James Joyce called him[58] when recommending a consultation to Nora, whose 'patients were sometimes so numerous that they overflowed into the hall' outside the ground-floor consulting rooms, in which he wrote prescriptions with the paper balanced on the palm of his hand and held Sunday dinners where medical, political and literary Dublin overlapped, was part of the very fabric of the late nineteenth and early twentieth-century city.[59] It is unfortunate that unlike Sir Dominic Corrigan, Sigerson's casebooks do not seem to have survived.[60]

In his first two decades in Dublin, Sigerson seems to have hovered politically between, on the one hand, the constitutional or 'Grattan nationalism' of The O'Donoghue and Sullivan and, on the other, the republicanism of the Irish Republican Brotherhood (IRB) and the Fenians.[61] He became a prolific contributor of political polemics to such publications as *The Irish People*, the journalistic mouthpiece of the IRB, *The Harp*, *The Shamrock* and *The Irishman*. According to Hyde, Sigerson contributed most of the leading articles for the latter during the period preceding the editorship of the enigmatic Richard Pigott.[62] Sigerson's 1867 *Irishman* editorial 'The holocaust', made a coruscating criticism of the executions of Allen, Larkin and O'Brien, and actually led to the imprisonment of Pigott as the journal's proprietor.[63] Brian Ó Cuiv has described Sigerson as an early proponent of policies that would be promulgated by Sinn Féin in the mid-1900s. Ó Cuiv attributes to Sigerson an 1868 editorial in *The Irishman* 'anticipating Arthur Griffith's abstentionist policy by nearly forty years', pointing to 'the futility of Irish representatives going to Westminster', and advocating the adoption of the tactics pursued so successfully against Austria by the Hungarians' in achieving dual-monarchy status with Austria after 1866.[64] Here, as in numerous other respects, Sigerson's advocacy of abstentionism followed the earlier lead of Young Ireland.

During this period Sigerson was also writing anonymously for English journals, contributing articles on the misgovernment of Ireland and what he presented as the valid reasons for the existence of the Fenian movement to *The Chronicle*. With Sigerson using the pseudonym 'An Ulsterman', these were later published in book-form as *Modern Ireland* (1868).[65] When the book went to a second edition in 1869, Sigerson used his own name. Read by Gladstone and the former Chief Secretary for Ireland, Lord Kimberley and by the English Catholic aristocrat-historian and Professor of Modern History in the University of Cambridge, Lord Acton, the book inspired Acton to invite Sigerson to write a series of articles on Irish land tenures for the *North British Review*.[66] This also subsequently appeared in book-form as the *History of Irish Land Tenures and Land Classes* (1871).[67]

By the 1880s, Sigerson was the recipient of a degree of establishment approbation. Now a Fellow of the Royal University, established under Disraeli during the previous decade, he corresponded with Acton on the subject of education. Having become involved with the situation of Fenian prisoners incarcerated in British jails, Sigerson was appointed to the Royal Prisons Commission in 1884, helping to prepare the ground for the Amnesty Act of 1885. Out of his involvements in this area came an 1890 book on the treatment of Irish political prisoners.[68] And following the precedent of Sir William Wilde's researches as Medical Commissioner to the Irish Census at the time of the Great Famine, Sigerson was sent in 1879 to the west of Ireland to investigate famine and the resulting diseases as Medical Commissioner to the Mansion House Relief Committee.

Meanwhile, on the literary front, the first edition of *The Bards of the Gael and the Gall*, Sigerson's much acclaimed anthology of Irish verse in English, appeared in 1897, dedicated by its editor to Charles Gavan Duffy and Douglas Hyde as representing 'the Gael and Gall respectively'.[69] Sigerson had become President of the Irish National Literary Society in 1893 and his address, chaired by Gavan Duffy at the society's inaugural meeting was later published as 'Irish literature, its origin, environment and influence' in the 1894 collection, *The Revival of Irish Literature*, alongside Gavan Duffy's essay of that title and Hyde's 'The necessity of de-anglicizing Ireland'.[70] Also at this time, writing as a social and political historian, Sigerson wrote the chapter on Grattan's Parliament for R. Barry O'Brien's *Two Centuries of Irish History* (1888, reprinted in 1907).[71] He later developed this into a book, published in 1918, in the context of the Sinn Féin landslide in the Westminster election of that year as *The Last Independent Parliament of Ireland*.[72]

Section II

By the 1900s, Sigerson was seen as the living embodiment of the School's direct continuity with its heroic early days under John Henry Newman's Rectorship. Since 1865, successive generations of students of the Catholic University School of Medicine had been coming to Newman House on St Stephen's Green to be taught the biological sciences by Sigerson. William Doolin recalled him as one of

> Newman's men ... the last surviving link with the early years of the School – our Professor of Biology, the venerable Dr. George Sigerson. Slow of gait and stooped of figure, with his great mane of silver-white hair and carefully tended moustache and beard, he looked every inch a descendent of the Vikings of old – a Sigur's son.[73]

Another old student remembered Sigerson as the 'handsome striking figure' who had so impressed Augustin Birrell, during his visit to the School as Chief Secretary for Ireland, that Birrell had declared that Sigerson should be retained after he had reached retirement age, 'as his appearance alone would shed lustre on the new university'.[74]

Within literary circles the veneration of Sigerson was equally if not more advanced by the early years of the new century. The May 1912 issue of J.S. Crone's *The Irish Book Lover*, for instance, noted favourably that

> a movement has been set on foot to acknowledge in a suitable manner the many services rendered to Irish literature by the veteran scholar, poet and scientist, Dr George Sigerson ... His *Poets and Poetry of Munster* has been before the public for half a century; his *Bards of the Gael and Gall* has recently been republished; his *Modern Ireland* and his standard work on *Land Tenures* did much to prepare the way for legislation that has effected such a change in the economic conditions of the country, and his famous article on 'The holocaust' ranks with Lady Wilde's 'Jacta alea est' as the finest specimen of Irish journalistic literature ever penned.[75]

A year later, the same column anticipated that 'The many friends and admirers of Dr George Sigerson will be delighted to hear that Mr John Lavery, the famous painter, has accepted the commission of the Sigerson Tribute Committee to paint the portrait of his fellow Ulsterman for the National Gallery of Ireland', and 'if funds permit, to also perpetuate the memory of the grand old man of Irish letters by a bust to be erected in some prominent place in the City of Dublin, the home of Dr Sigerson during the last fifty years'.[76] In April 1913, *The Irish Review: A Monthly Magazine of Irish Literature, Art and Science*, founded by Padraic and Mary Colum with Thomas MacDonagh, James Stephens and Thomas Houston, carried as its frontispiece a monochrome reproduction of

Lavery's *Portrait of Dr Sigerson*.[77] Meanwhile in the 'Editor's Gossip' for the August–September 1920 *Irish Book Lover*, J.S. Crone recalled a recent visit to Dublin where his first call was to 'Dr Sigerson, the grand old doyen of Irish literature' who proudly showed him 'a beautiful death mask of Charles Kickham' he had recently acquired.[78]

When Sigerson's *The Last Independent Parliament of Ireland* was republished as a book in its own right in 1918, the author of the May *Irish Book Lover* review of that year eulogised Sigerson as someone 'who has been in the forefront of the many Irish movements, literary and political, the friend and adviser of men who have played prominent parts during the last sixty years'. And when Seanad Eireann sat for the first time on 11 November 1922, it honoured Sigerson as its oldest member by electing him Chairman for that day. In seconding this temporary appointment, Sigerson's fellow-historian Alice Stopford Green referred to the 'honour' of 'so learned and faithful an historian of Ireland' taking the Chair for that day.[79]

After Sigerson's death in 1925, Douglas Hyde wrote a 'Memorial preface' to the new edition of Sigerson's *Bards of the Gael and the Gall* published that year.[80] In his opening paragraph, Hyde observed that

The remarkable man who passed away from amongst us on the 17 February, 1925, has left a gap in our midst that cannot be filled. He was as it were a giant oak which had seen generations of lesser trees grow around it and pass away, and its own fall in the fullness of years has left vacant the space which had been adorned by its stately presence. The most notable link that connected the Ireland of Penal times, and certainly with the Ireland of nearly three-quarters of a century ago, has at last snapped. Full of old age and well-deserved honours and reverence and love this outstanding figure has departed from amongst us.

While Hyde found it difficult to properly appraise one whose attainments had been so many and various, what he found most striking about Sigerson was

his own career, during which he worked incessantly, never wavering in his affection for things Irish, never halting in giving his allegiance and best service to the cause of mankind and of his own country ... As an Irish scholar he was the last link that connected us with the era of O'Donovan and O'Curry, and one of the last that connected us with the men of '48, with Kickham and with Mitchel.

He had known them all, shared their counsels and aspirations, befriended and sheltered many of them, and could tell of them from the intimacy of close association in a way that was the privilege of no other living person. He was not the child of any one province; he was all-Ireland, and one might say cosmopolitan. Born and reared near Strabane, but with a Kerry ancestry, educated partly in Galway, partly in Cork, and later in Paris, he typified all that was best and broadest and sanest in our race...He died in his ninetieth year in the full enjoyment of his faculties, and Ireland will not forget him and cannot replace him.[81]

Section III

Hyde's confidence in Sigerson's survival in Irish national memory would appear to have been misplaced when seen from the context of today. Although there are short entries on Sigerson in the 1928 and 1978 dictionaries of Irish biography by Crone and Henry Boylan respectively, in Kate Newman's (1994) *Dictionary of Ulster Biography* and in Robert Welch's *Concise Oxford Companion to Irish Literature*,[82] there is no entry as yet for Sigerson in the *Oxford Dictionary of National Biography*.[83] Sigerson is also absent from Sean Connolly's *Oxford Companion to Irish History*, while the entry by Brian Cliff in the *Encyclopaedia of Ireland* merely notes that, 'Apart from influential anthologies and translations of Irish poetry, his main intellectual contribution was probably his continual opposition, from within cultural nationalism, to theories of national purity'.[84]

On the scientific front, meanwhile, Sigerson is briefly memorialised in Ray Desmond's *Dictionary of Irish and British Botanists and Horticulturalists* (1994).[85] His prominent public profile between the 1860s and the 1900s, as a journalist and editor in the advanced nationalist press and as a historian and polemicist, whose voice was noted in British establishment circles by Gladstone and Lord Acton amongst others, has been largely neglected by the historians of this period. John Powell and Padraic Kennedy's study of the influence of Sigerson's *Modern Ireland* (1868) on the Liberal Minister Lord Kimberley and subsequently on Gladstone himself is, however, a notable exception to this neglect.[86] Otherwise, Sigerson gets what can best be described as a walk-on part in most accounts of later Victorian Irish nationalism.[87]

The three most recent overall studies of the Revival, moreover, make only scant reference to Sigerson. In P.J. Matthews' 2003 book, *Revival: The Abbey Theatre, Sinn Féin, the Gaelic League and the Co-operative*

Movement, and Sinead Garrigan Mattar's 2004 study, *Primitivism, Science and the Irish Revival*, Sigerson is awarded one index entry apiece, though in Gregory Castle's 2001 work, *Modernism and the Celtic Revival*, he scores slightly better.[88] This is rather surprising given, as Nicholas Allen's review of Garrigan Mattar in the February 2005 *Irish Studies Review* notes, 'the recent trend in inter-disciplinary study of the Revival as a trans-cultural moment of late-nineteenth and early-twentieth century Ireland'. Allen, indeed, sees the originality of Garrigan Mattar's work as lying in her 'attention to primitivism and science as major discourses of the Revival' and her presentation of the Revival as 'a movement of modernity'.[89] If, however, Sigerson's working connections with French psychiatry, neurology and the Parisian human sciences, together with his Dublin academic and scholarly career as a man of science and clinician, are taken alongside his contributions to the canon of Irish nationalism and to the Revival, then the sum total of these interventions arguably provides a most apposite exemplification of the very particular ways in which the Revival constituted 'a movement of modernity'.

There are two notable exceptions to the general latter-day neglect of Sigerson. These are in two discrete areas of scholarship: J.B. Lyons' work on Irish medical history, on the one hand, and the interdisciplinary critical analyses of the Field Day cohort in Irish studies, on the other. J.B. Lyons was the grand old man of the history of medicine in Ireland. Manifesting the finest characteristics of the 'amateur' tradition, he was both a consultant practitioner and a prolific author. Amongst the numerous essays and books constituting his medical history *oeuvre* are a series of essays on Sigerson presenting him as a doctor, a man of science and a polymath.[90] The 1997 'George Sigerson: Charcot's English translator', published in the *Journal of the History of the Neurosciences*, summarises his work in the medical and literary fields, foregrounding him as 'Dublin's first neurologist', and placing particular emphasis on his relationship with Charcot. Lyons ascribes Sigerson's lack of a hospital appointment to his involvement in 'political journalism' and notes that the first post – as a lecturer in botany at the Catholic University – was in an 'unendowed and struggling institution' and that, notwithstanding his later professorial elevations, he had to rely on his practice to produce an adequate living. Lyons suggests that the MacMahon sword episode may have given Sigerson the opportunity not just to reacquaint himself with the Paris of his schooldays at the Collège St Joseph but also to start to explore the city's medical world. In terms of Sigerson's associations with Duchenne, Charcot *et al.*,

Lyons points out that Charcot and Sigerson belonged to the same generational cohort. Charcot's *aggrégation* in 1860 was more or less contemporary with Sigerson taking his M.D. The former's initial appointment as physician to the Salpêtrière was around the time Sigerson took up his first Catholic University teaching post. Lyons speculates that by getting to know Charcot early in his career was 'at a time propitious for the kindling of a friendship'.[91] Lyons also uncovered a series of letters between Sigerson and Charcot in the National Library of Ireland.

Lyons later qualified his 1997 designation of Sigerson as 'Dublin's first neurologist',[92] on the basis of Sigerson's association with Duchenne, Charcot and the New Sydenham Society translation of the latter, by acknowledging that, by contrast with Francis Purser (1876–1934), 'Sigerson lacked a hospital appointment; his academic attachment was in botany and zoology, and he was deeply involved in history and literature.'[93] Although Francis Purser had connections with the arts – he was the nephew of the artist Sarah Purser, the first female member of the Royal Hibernian Academy and Mespil House *salonière* – in his career he was very much the Irish embodiment of the scientific role-model for elite clinicians that was increasingly in the ascendant by the 1900s.[94]

Purser, like Sigerson, had studied neurology on the continent (Germany in his case) and also at the National Hospital for Nervous Diseases in Queen's Square, London. Unlike Sigerson, however, Purser did not confine his Dublin practice of neurology to the private sphere of consultancy. He held appointments as a general physician at Mercer's Hospital and as a senior physician at the Richmond, Whitworth, and Hardwick Hospitals. He served as consulting neurologist to the British forces in Ireland, publishing several significant contributions to the medical debate on shell-shock during the First World War. In 1926, he was appointed to a personal chair at Dublin University.[95]

In the selections from Sigerson's political writings, poetry, translations, and anthologies, made by Seamus Deane in the *Field Day Anthology* (1991), Sigerson's centrality to nationalist polemics from the 1860s to the Revival, and to accompanying debates about Irish identity and tradition, is emphasised. Deane sees Sigerson as seeking to transcend established stereotypes of national character through his emphasis on the multi-vocal inclusiveness of Irishness, the Gall alongside the Gael.[96] Other work by Deane, Luke Gibbons, David Dwan and Terry Eagleton, Deane's Irish studies associates in the Field Day cohort, suggests or provides analogies for contextualising Sigerson's life and career. In particular, their work points to ways of seeing how Sigerson's twinning of a commitment to a post-Enlightenment modernising meliorism, enhanced in his case by medicine, with an invocation of, and sense of continuity with, the Gaelic past was in turn characteristic of the broader sweep of Irish intellectual culture from the middle of the eighteenth century onwards.

Luke Gibbons has delineated such a conjuncture in the Roscommon circle around Charles O'Connor of Ballanegare in the 1780s and 1790s, where gentry patronage of residual Gaelic culture co-existed with enlightened plans for the development of local industry, a conjuncture that was dissipated in the counter-revolutionary terror that followed the failure of the United Irishmen in 1798. Gibbons sees the 'radical memory' of this cultural circle surviving but transforming, through the perceptions inculcated by the culture of Gothic literature and art in the early nineteenth century, to become part of 'a phantom public sphere haunted by fear, terror and the dark side of civility'.[97] In *Strange Country: Modernity and Nationhood in Irish Writing since 1790* (1997), Seamus Deane has brought out the persistently different ways in which, by sharp contrast with Scotland, the legacy of Enlightenment had been digested in Ireland.[98] One facet of the Enlightenment in particular, the civic republican tradition, has been seen by David Dwan as providing the most useful 'backdrop' against which to view the politics of Young Ireland,[99] a movement with whose residual cohorts in Cork, and later Dublin, Sigerson was intimately connected from at least his days as a medical student at Queen's College, Cork, in the second half of the 1850s.

Dwan 'challenges' attempts by political theorists to differentiate 'the language of civic republicanism or patriotism from the discourses of nineteenth-century nationalism', seeing Young Ireland in the 1840s as a group of 'self-confessed nationalists ... [who] also invoked the languages and ideals of an earlier [eighteenth-century] patriot tradition', problematic as 'this republican advocacy' may have been when 'articulated in a mid nineteenth-century context' of 'a discursively ascendant Smithian political economy'.[100] He has subsequently enlarged on this theme at greater length in his 2008 book *The Great Community* in which he reappraises Irish cultural nationalism from Young Ireland to the early W.B. Yeats, presenting it as 'an ambitious attempt to recover an ancient ideal of citizenship for a modern democratic age'.[101]

In *Heathcliff and the Great Hunger* (1995), Terry Eagleton had already noted how 'the old and the new continued to form strange conjunctures' in a post-Famine Ireland where Irish nationalism and cultural modernism alike were the products of an 'archaic avant-garde'.[102] The savants who formed a significant component of this 'archaic avant-garde' with its 'bizarre blending'[103] of technology and the gothic, are epitomised for the Eagleton of *Scholars and Rebels in Nineteenth-Century Ireland* (1999) by George Sigerson.[104] For Eagleton, Sigerson is 'The link between the medical community of Victorian Dublin and the Irish Revival'. He celebrates Sigerson as an author whose range extended to 'paralysis and St. Patrick, Norse Ireland and the advantages of ambidexterity'.[105] If 'The range of the work, from law, emigration and education to medicine and the Irish convict system, reflects the familiar encylopaedism of the Irish Victorian scholar', infused with 'cultural supremacism', Eagleton finds Sigerson nonetheless 'a genuinely literary cosmopolitan, fascinated by cross-cultural fertilisations'.[106] Sigerson's utility for facilitating the 'cross-cultural fertilizations' of others is demonstrated in Luke Gibbons's 1997 *Irish University Review* article 'Some hysterical hatred: history, hysteria and the Literary Revival'.[107] Here Sigerson doubles up, featuring both as the translator of Charcot and as the author who explains to an English audience how the land issue was understood in Ireland. In Sigerson's translation, Charcot's *Diseases of the Nervous System* suggestively provides Gibbons with a model of hysteria that can be used to understand what Gibbons saw as the hysterical underpinnings of Irish memory in writings of the Revival, like Bram Stoker's *The Snake's Pass* (1890), a novel whose narrative centres around a moving bog in the west of Ireland that was rumoured to contain buried French treasure from 1798.[108]

Conclusion

As an eminent medical man, a man of science and a public intellectual, George Sigerson can be seen as belonging to an Irish tradition that also includes Sylvester O'Halloran[109] and Sir William Wilde.[110] Like O'Halloran, Sigerson trained on the continent and introduced continental medical innovations into Ireland. Like Wilde, he combined a specialist private practice in the key vicinity of the Merrion Square area of Dublin, with significant and multifarious involvements in the city's metropolitan culture and a number of its key institutions, including the Royal Irish Academy. Also like Wilde, Sigerson had an international reputation in both the bio-medical and human sciences with concomitant membership of elite European institutions. But unlike Wilde, he founded no hospital for his specialism, nor was his medical career marked by a consultancy in one of the great Dublin hospitals. His investigations of famine conditions in 1879 as Medical Commissioner for the Mansion House Relief Committee pale in comparison to Wilde's work as Commissioner for the Irish Census over four decades.[111] Sigerson did, however, enjoy a distinguished academic career as the holder of two consecutive professorial chairs at the Catholic University School of Medicine. And across the broad canvas of Victorian Dublin, his political involvements were more radical and more extensive than those of Wilde. As a medical man and public intellectual, R.R. Madden rather than Wilde would be a more appropriate point of comparison with Sigerson in this respect.[112]

Like the lives and careers of Sylvester O'Halloran, Richard Robert Madden and Sir William Wilde, that of George Sigerson can be seen as part of a specifically Irish conversation between science, modernity and national identity, which differs notably from the equivalent and better-known English conversation.[113] This in turn provides an Irish dimension to the historiographical debate about national styles of science.[114] It illustrates a particular relationship between medicine, culture and politics that can be found in Ireland during the eighteenth, nineteenth and early twentieth centuries. This relationship casts new light on the wider history of the island, by demonstrating how medical men contributed to the construction of an emergent national intelligentsia, and by shedding light on the wider nature of the Irish intellectual class that came into being during this period. In the case of George Sigerson, we see a public intellectual engaged at various levels of culture and politics with questions of national and cultural identity; a major participant in the Revival who was both a clinician and a man of science with an international reputation.

Lest Sigerson's self-fashioning through these myriad interventions in the public sphere seem like a late Victorian Hibernian not-fully-professionalised throwback to Robert M. Young's intellectual 'common context' of the earlier nineteenth century,[115] however, it should be noted that his career displays many of the characteristics of the Victorian 'man of science', as adumbrated in Paul White's seminal study of T.H. Huxley.[116] White sees the upwardly mobile Huxley, who was the son of an elementary schoolmaster, as negotiating a gentlemanly scientific identity for himself, which was distinct from the later model of the laboratory-based, discursively specialised professional 'scientist'. As a scientific practitioner, White's Huxley negotiated and articulated a public and moral role for himself as a scientific practitioner who actively engaged with the high culture of the society in which he lived, intervening in contemporary debates in politics, religion, philosophy, economics and literature. In a seminal article,[117] Christopher Lawrence has demonstrated the persistence within the British medical élite between the middle of the nineteenth century and the beginning of the First World War, of significant pockets of a gentlemanly, culturally conservative culture of care. Sceptical of claims being made by some clinicians for a professional and exclusively 'scientific' medical identity, and displaying pronounced continuities with the Coleridgean and Anglican medical establishment of the 1820s and 30s,[118] these eminent practitioners significantly delayed the eventual triumph of a professional and exclusively scientific culture of care.[119]

Furthermore, and directly within the area of Sigerson's own medical specialism, the career of the English neurologist Sir Henry Head (1861–1940), 26 years younger than Sigerson and a published poet who was closely associated with Thomas Hardy and Siegfried Sassoon, shows that in England a cutting-edge research career as a medical specialist was still culturally compatible with a literary profile in the early twentieth century.[120] Indeed, Head's biographer, L.S. Jacyna argues that Head's ground-breaking work on how language and the brain were related through the nervous system has to be understood in the context of contemporary debates within modernist philosophy and the arts about perception and language. Jacyna's Henry Head was 'an aspirant poet and man of letters as well as a rising figure in the world of medicine', self-consciously bohemian and anti-bourgeois, who looked to the countryside as relief from the metropolis like a J.C. Squire or Edmund Blunden. Jacyna sees Head's work as spanning the cognitive modernism, manifest in his work on the physiology of sensation, and the aesthetic modernism of his poetry and avant-garde inclinations in music, drama and fiction.[121] The Sigerson of the Revival, the poet and nationalist polemicist with his radical concerns about the iniquities of Irish land tenures, the condition of Irish political prisoners, and the preservation of Irish language and its literature is also, like Henry Head, recognisably a figure of the European *fin-de-siècle*, a European identity reinforced in Sigerson's case through his connections with French medicine and its culture of care.

Alongside these English and European parallels, however, Sigerson should essentially be seen in terms of the particular Irish lineage he shared with O'Halloran, Madden and Wilde. Where Henry Head sought to combine cognitive and aesthetic modernism, Sigerson's commitment was to a modernity capable of carrying with it the Gaelic past. Unlike Head, however, Sigerson still awaits an appraisal that integrates his work in the bio-medical sciences with his role in the Revival and contemporary nationalist polemic. Gregory Castle and Sinead Garrigan Mattar have demonstrated how Yeats and Synge deployed late-nineteenth century anthropology; Castle considers the imperial anthropology of A.C. Haddon, while Garrigan Mattar looks at the new (French) science of 'Celtology'.[122] But of all the major figures of the Revival, only Sigerson actually practised science and enjoyed an international reputation as a man of science. For all of these reasons, a new assessment of Sigerson is well overdue.

Notes

1 The most detailed account of Sigerson's life and career is Norah Fahie, 'Dr George Sigerson', *Dublin Historical Record*, 38 (December 1984–September 1985), pp. 53–60. See also J.B. Lyons, 'Medicine and literature in Ireland', *Journal of the Irish Colleges of Physicians and Surgeons*, 3:1 (July 1973), pp. 3–9; idem., 'George Sigerson (1836–1925)', in idem., *Brief Lives of Irish Doctors* (Dublin, 1978), pp. xx–xxii; idem., 'Neurology in Hibernia and in Hibernia Major', *Cogito*, 1 April 1989, pp. 73–5; idem., 'George Sigerson: Charcot's English translator', *Journal of the History of the Neurosciences*, 6:1 (April 1997), pp. 50–60. The entry on Sigerson in the forthcoming *Dictionary of Irish Biography* is also by Lyons. See also Diarmuid Breathnach and Máire Ni Mhurchu, 'Sigerson, George, 1836–1925', in idem., *1882–1982 Beathaisnéis A hAon* (Dublin, 1986), pp. 112–13; 'Sigerson, George', in Alan R. Eager, *A Guide to Irish Bibliographical Material* (London, 1980), pp. 1123–4. To date there is no entry for Sigerson in the *Oxford Dictionary of National Biography*.
2 See http://cao.gaa.ie/sigerson.htm. For Sigerson and Gaelic games in his birthplace see 'Strabane Sigersons', http://ulster.gaa.ie/tyrone%20clubs/strabane.html. See also Joseph Martin, *The G.A.A. in Tyrone: The Long Road to Victory* (Omagh, 2003).
3 Dora Sigerson Shorter, *The Tricolour: Poems of the Irish Revolution* (Dublin, 1922). For Dora Sigerson Shorter see Evelyn A. Hanley, 'Dora Sigerson Shorter: Late Victorian romantic', *Victorian Poetry*, 3:4 (Autumn 1965), pp. 223–4.
4 Lyons refers to 'inherited ... lands and a spade mill at Holy Hill', in Lyons, 'Medicine and literature in Ireland', p. 5.
5 Unless otherwise specified, the biographical details in the pages which follow are largely taken from Fahie, 'Dr George Sigerson'.
6 For the politics of the north-west see Brendan MacSuibhne, '"Patriot paddies": The volunteers and Irish identity in Northwest Ulster, 1778–86' (Unpublished Ph.D. thesis: Carnegie Mellon University, 1998).
7 Douglas Hyde, 'George Sigerson: Born on January 11 1836 – died February 17 1925. A memorial preface', in George Sigerson, *Ballads of the Gael and Gall: Examples of the Poetic Literature of Erin; done into English after the Metres and Modes of the Gael* (Dublin, 1925), p. v.
8 Hyde attributes the move to Cork as a result of Sigerson's having contracted typhoid fever in Galway. Ibid., p. 50.
9 Lyons sees Sigerson's decision to take up the university study of Irish alongside medicine to the influence of 'a group of nationalists', presumably the circle around the Varians. See Lyons, 'Charcot's English translator'.
10 Ralph Varian (ed.), *The Harp of Erin: A Book of Ballad – Poetry and Native Song* (Dublin, 1869).
11 Lyons, 'Charcot's English translator', p. 51.

12 William Doolin, 'The Catholic University School of Medicine (1855–1909)', in Michael Tierney (ed.), *Struggle with Fortune: A Miscellany for the Centenary of the Catholic University of Ireland* (Dublin, 1954), p. 73.

13 George Sigerson, 'Additions to the flora of the tenth botanical district, Ireland' and 'Discovery of fish-remains in the alluvial clay of the River Foyle', *Proceedings of the Royal Irish Academy*, Ser. 2:1. Science (1869–74), pp. 192–8, 212–14.

14 George Sigerson, 'On heat as a factor in vital action (so called)', *Proceedings of the Royal Irish Academy*, Ser. 2:2. Science (1875–77), pp. 1–6.

15 For John Tyndall see W.H. Brock, N.D. Macmillan and R.C. Mollan (eds), *John Tyndall: Essays on a Natural Philosopher* (Dublin, 1981). For the X-Club see R.M. McCloud, 'The X-Club', *Notes and Records of the Royal Society*, 24 (1969–70), pp. 305–22; Ruth Barton, '"An influential set of chaps": The X-Club and Royal Society politics, 1864–1885', *British Journal for the History of Science*, 23 (1990), pp. 53–81; idem., '"Huxley, Lubbock, and half a dozen others": Professionals and gentlemen in the formation of the X Club, 1851–1864', *Isis*, 89:3 (1998), pp. 410–44.

16 Sigerson to Darwin, 8 July 1863, in Frederick Burkhardt and Sydney Smith (eds), *The Correspondence of Charles Darwin*, 11 (1863) (Cambridge, 1999), p. 529.

17 Fahie, 'Dr George Sigerson', p. 56.

18 J.R. Lucas, 'Wilberforce and Huxley: A legendary encounter', *Historical Journal*, 22:2 (1979), pp. 313–30.

19 George Sigerson, 'Considerations of the structural and acquisitional elements in dextral pre-eminence, with conclusions as to the ambidexterity of primeval man', *Proceedings of the Royal Irish Academy*, Ser. 2:4. Science (1884), pp. 38–50. For handedness see L.L. Harris, 'On the evolution of handedness: A speculative analysis of Darwin's views and a review of early studies of handedness in "the nearest allies of man"', *Brain and Language*, 73:2 (June 2000), pp. 132–88. For the nineteenth-century debate on human evolution see J.W. Burrow, *Evolution and Society: A Study in Social Theory* (Cambridge, 1966); Peter J. Bowler, *Theories of Human Evolution: A Century of Debate, 1844–1944* (Baltimore, 1986).

20 George Sigerson, 'Genesis and evolution', *New Ireland Review*, 1 (1894), 18–26, 87–93.

21 For the intertwined histories of the Catholic University, the Royal University and University College Dublin during this period see Tierney (ed.), *Struggle with Fortune*; Donal McCartney, *UCD: A National Idea: The History of University College, Dublin* (Dublin, 1999); John Coolahan, 'From Royal University to National University, 1879–1908', in Tom Dunne (ed.), *The National University of Ireland: Centenary Essays* (Dublin, 2008), pp. 3–19.

22 *Royal Commission on University Education (Ireland)*, 31, *Twentieth Report of the Royal University of Ireland* (Sessional papers Cd. 1460), p. 92. Delaney concluded his statement by affirming that 'There was no ground for the allegation – none whatever', p. 78, Cd. 1205A. Delaney had earlier given 'the most point blank statement again to a grievous misstatement made … by a member of the House of Commons … that a certain former Professor of Stephen's-green had been dismissed on account of his teaching in Biology'. There was absolutely no foundation for that statement, which appeared in *The Times*.

23 For the Independent Party see John Whyte, *The Independent Irish Party* (London, 1958); Stephen R. Knowlton, *Popular Politics and the Irish Catholic Church: The Rise and Fall of the Irish Independent Party, 1850–1857* (New York, 1991).

24 'An Irish sword for Marshal MacMahon', *Irish Sword*, 4:17 (1980).

25 Lyons, 'Charcot's English translator', p. 52.

26 For the Paris clinical school see Toby Gelfand, *Professionalizing Modern Medicine: Paris Surgeons and Medical Science and Institutions in the Eighteenth Century* (Westport, Conn., and London, 1980). For the development of scientific medicine in the nineteenth century see W.F. Bynum, *Science and the Practice of Medicine in the Nineteenth Century* (Cambridge, 1994).

27 'By the end of 1859 Sigerson was in Dublin, hoping to specialize in neurology ...'. Fahie, 'Dr George Sigerson', p. 54. For neurology see Frank Clifford Rose and W.F. Bynum, *Historical Aspects of the Neurosciences* (New York, 1982); W.F. Bynum, 'The nervous patient in eighteenth and nineteenth-century Britain: The psychiatric origins of British neurology', in W.F. Bynum, Roy Porter and Michael Shepherd (eds), *The Anatomy of Madness: People and Ideas 1* (London, 1985), pp. 89–102; Mark Micale, 'Hysteria and its historiography', *History of Science*, 27 (1998), pp. 233–61; Janet Oppenheim, *'Shattered nerves': Doctors, Patients and Depression in Victorian England* (New York, 1991).

28 Among Louis-Antoine Ranvier's publications were, with V. Cornil, *Manuel d'histologie pathologique* (Paris, 1869); *Traite technique d'histologie* (Paris, 1875); with E. Weber, *Leçons sur l'histologie du système nerveux* (Paris, 1878).

29 For Ranvier see 'John Gach Books, Inc., French neuroscience and medicine', http://www.gach.com/Gach/1580-03.htm, 06/03/2009.

30 Benjamin Ball, *Du délire des persecutions ou Maladie de Charles Lasegue* (Paris, 1890). For Ball, see http://www.artandmedicine.com/biblio/authors/revue/Ball.html, 06/03/2009.

31 Jean-Martin Charcot and Benjamin Ball, *Leçons cliniques sur les maladies des vieillards et les maladies chroniques* (Paris, 1874).

32 For Charcot see Christopher G. Goetz, Michel Bonduelle and Toby Gelfand, *Charcot: Constructing Neurology* (New York and Oxford, 1995).

33 For Duchenne see J. L'Hermitte, 'Duchenne de Boulogne et son temps', *Bulletin de l'Academie de Médécine (Paris)*, 130 (1946), pp. 745–55; V. Dubowitz, 'History of muscle disease', in Rose and Bynum (eds), *Historical Aspects of the Neurosciences*, pp. 13–22; A.E.H. and M.L.H. Emory, *The History of a Genetic Disease: Duchenne Muscular Dystrophy or Meryon Disease* (London, 1985); http://www.artandmedicine.com/biblio/authors/revue/Duchenne.html, 06/03/2009.

34 Goetz, Bonduelle and Gelfand refer to Sigerson, in the context of the first New Sydenham Society translation, as 'an Irish physician and disciple of Duchenne de Boulogne'. Goetz, Bonduelle and Gelfand, *Charcot*, p. 220.

35 Fahie, 'Dr George Sigerson', p. 55, refers to how after attending the 1879 British Medical Association annual meeting held in Cork, Charcot travelled around Ireland in Sigerson's company, dined in Sigerson's Dublin home and later invited Sigerson and his two daughters, Dora and Hester to his home in Neuilly.

36 Lyons, 'Charcot's English translator', p. 52, with reference to Goetz, Bonduelle and Gelfand, *Charcot*, p. 54.

37   J.B. Lyons, 'Charcot's English translator', p. 54, refers to *Le Progrès Médical* 'a journal closely associated with the Salpêtrière'.
38   George Sigerson, 'On alternate paralyses', *Dublin Journal of Medical Science*, 65 (1878), pp. 97–125.
39   Jean Martin Charcot, *Leçons sur les maladies du système nerveux faites à Salpêtrière* (Paris, 1875).
40   J.M. Charcot, *Lectures on the Diseases of the Nervous System delivered at La Salpêtrière*, trans. by George Sigerson and Thomas Savill (3 vols, London, 1877, 1881 and 1889). Savill translated the 1889 volume. For the New Sydenham Society (1858–1911) see G.C. Meynell, *The Two Sydenham Societies: A History and Bibliography of the Medical Classics Published by the Sydenham Society and the New Sydenham Society (1844–1911)* (Acrise, Kent, 1985).
41   George Sigerson, 'A lecture on certain phenomena of hysteria major delivered at La Salpêtrière, by Professor Charcot', *British Medical Journal*, 2 (1878), pp. 789–91.
42   J.M. Charcot, *Poliklinische Vorträge/von J.M. Charcot. 1. Bd. Schuljahr 1887–88 ubersetzt von Sigmunnd Freud* (Leipzig, 1892–95). As with the third volume of the English translation where a different translator was used rather than Sigerson, in the further German volume of the *Leçons du mardi à la Salpêtrière:policliniques*, covering the 1888–89 academic year, the translation was by Max Kehane, not Freud. There was no English translation of the *Leçons du mardi*, until Christopher Goetz's centenary selection from 1887–88 as *Charcot the Clinician: The Tuesday Lessons. Excerpts from Nine Case Presentations on General Neurology delivered at the Salpêtrière, Hospital in 1887–88 by Jean-Martin Charcot*, trans. with commentary by Christopher G. Goetz (New York, 1987). For Freud and Charcot see Kenneth Levin, *Freud's Early Psychology of the Neurosciences: A Historical Perspective* (Haassocks: Sussex, 1978), pp. 42–63; Peter Gay, *Freud: A Life for our Time* (London, 1988), pp. 48–53; Toby Gelfand, 'Charcot's response to Freud's rebellion', *Journal of the History of Ideas*, 50:2 (April–June, 1989), pp. 293–307.
43   Douglas Hyde, *The Revival of Irish Literature* (London, 1894), p. ix.
44   For a discussion of Sigerson's Parisian mentors see Fahie, 'Dr George Sigerson', pp. 55–6.
45   See Lyons, 'Charcot's English translator', p. 54.
46   George Sigerson, 'An examination into certain recently reported phenomenon in connexion with hystero-epilepsy' and 'Influence of solenoids on the nervous system', *British Medical Journal*, 1 (1879), pp. 143–5, 620–1. I am grateful to J.B. Lyons, 'Charcot's English translator' for this reference.
47   For Paul Broca and the Anthropological Society of Paris see L.S. Jacyna, *Lost Words: Narratives of Language and the Brain, 1825–1926* (Princeton and Oxford, 2000).
48   Sigerson, 'Genesis and evolution'.
49   For Coffey see Doolin, 'The Catholic University School of Medicine'.
50   For the specifically Irish dimensions of this critique see Greta Jones, 'Catholicism, nationalism and science', *Irish University Review*, 20 (Winter/Spring 1997), pp. 47–61.

51 For French reservations about Darwinism see Robert E. Stebbins, 'France', in Thomas E. Glick (ed.), *The Comparative Reception of Darwinism* (Austin and London, 1974), pp. 117–63. But see also, for a general *fin de siècle* 'eclipse of Darwinism', Peter J. Bowler, *The Eclipse of Darwinism* (Baltimore and London, 1983).

52 For sectarianism and nineteenth-century Irish hospital appointments see Laurence M. Geary, *Medicine and Charity in Ireland, 1718–1851* (Dublin, 2004).

53 For Hall's tribulations in London see Bynum, 'The nervous patient', p. 94.

54 For the general hostility towards medical specialism of the nineteenth-century British medical establishment see Lindsay Granshaw, '"Fame and fortune by means of bricks and mortar": The medical profession and specialist hospitals in Britain, 1800–1948', in Lindsay Granshaw and Roy Porter (eds), *The Hospital in History* (London, 1989), pp. 199–220. For Sir William Wilde and his specialist practice in Merrion Square and in the infirmary he established as St Mark's Hospital for the Diseases of the Eye and Ear see T.G. Wilson, *Victorian Doctor: Being the Life of Sir William Wilde* (London, 1942); James McGeachie, '"Normal" developments in an "abnormal" place: Sir William Wilde and the Irish school of medicine', in Greta Jones and Elizabeth Malcolm (eds), *Medicine, Disease and the State in Ireland, 1650–1940* (Cork, 1999), pp. 85–101 and idem., 'Wilde, Sir William, 1815–1876', *Oxford Dictionary of National Biography* (Oxford, 2004).

55 For Merrion Square and the Dublin medical establishment see McGeachie, '"Normal" developments'.

56 Lyons, 'Charcot's English translator', p. 51.

57 See Lyons, *Brief Lives*, for Sigerson and his patients.

58 Ibid.

59 Fahie, 'Dr George Sigerson', p. 58.

60 For Corrigan and his casebooks see Eoin O'Brien, *Conscience and Conflict: A Biography of Sir Dominic Corrigan 1802–1880* (Dublin, 1983).

61 For Sigerson as a 'Grattan nationalist' see Richard Pigott, *Recollections of an Irish National Journalist* (Dublin and London, 1882).

62 For Pigott see Marie-Louise Legg, 'Pigott, Richard (1828–1889)', *Oxford Dictionary of National Biography* [http://www.oxforddnb.com/view/article/22255, accessed 8 November 2006]; Margaret O'Callaghan, 'Richard Pigott, the fringe Fenian press and the politics of Irish nationalist transition to Par-nellism', in James McConnell (ed.), *The Black Hand of Irish Republicanism: Fenianism in Modern Ireland* (Dublin, 2009), pp. 149–59.

63 For this episode see Legg, 'Pigott, Richard'.

64 Brian Ó Cuiv, 'The Gaelic cultural movements and the new nationalism', in Kevin B. Nowlan (ed.), *The Making of 1916: Studies in the History of the Rising* (Dublin, 1969), p. 13.

65 An Ulsterman, *Modern Ireland: Its Vital Questions, Secret Societies and Government, by an Ulsterman* (London, 1868).

66 John Powell and Padraic Kennedy, 'Lord Kimberley and the foundation of Liberal Irish policy: Annotations to George Sigerson's *Modern Ireland: Vital Questions, Secret Societies and Government (1868)*', 'Selected Documents, 48', *Irish Historical Studies*, 31 (1998), pp. 91–114.

67 George Sigerson, *History of Irish Land Tenures and Land Classes of Ireland. With an Account of the Various Secret Agrarian Confederacies* (Dublin, 1871).

68   George Sigerson, *Political Prisoners at Home and Abroad* (London, 1890).
69   George Sigerson, *The Bards of the Gael and the Gall* (Dublin, 1897).
70   Charles Gavan Duffy, George Sigerson and Douglas Hyde, *The Revival of Irish Literature* (London, 1894).
71   R. Barry O' Brien (ed.), *Two Centuries of Irish History, 1691–1870* (London, 1907).
72   George Sigerson, *The Last Independent Parliament of Ireland, with an Account of the Survival of the Nation and its Lifework* (Dublin, 1918).
73   Doolin, 'The Catholic University School of Medicine', p. 73.
74   F.O.C. Meenan, *Cecilia Street: The Catholic University School of Medicine 1855–1931* (Dublin, 1987), pp. 88–9.
75   Anon., 'Gossip', *The Irish Book Lover*, 3 (1912), p. 167.
76   Ibid., 4 (1913), p. 31.
77   'Portrait of Dr. Sigerson by John Lavery, R.H.A., A.R.A., R.S.A.', *The Irish Review*, 3 (1913), facing p. 57; J.S. Crone, 'Editor's gossip', *The Irish Book Lover*, 10 (1920), p. x.
78   *Irish Book Lover*, 9 (1918), p. 136.
79   'Seanad Éireann, 1 (11 December 1922) Election of temporary Chairman', <http://historical-debates.oireachtas.ie>.
80   Ibid., pp. v–vi, xxii.
81   Douglas Hyde, 'George Sigerson: A memorial preface', in George Sigerson, *Ballads of the Gael and Gall* (3rd edn, Dublin, 1925), pp. v–vi.
82   J.S. Crone, *Dictionary of Irish Biography* (Dublin, 1928); Henry Boylan, *Dictionary of Irish Biography* (Dublin, 1978); Kate Newman (ed.), *Dictionary of Ulster Biography* (Belfast, 1994); 'Sigerson, George', in R. Welch (ed.), *The Concise Oxford Companion to Irish Literature* (Oxford, 2000).
83   J.B. Lyons, 'Sigerson, George', in J. McGuire and J. Quinn (eds), *The Dictionary of Irish Biography*, 8 (Cambridge, 2009), pp. 945–6.
84   S.J. Connolly (ed.), *Oxford Companion to Irish History* (Oxford, 2002, 2nd edition 2007); Brian Cliff, 'Sigerson, George', in Brian Lalor (ed.), *Encyclopedia of Ireland* (Dublin, 2003), p. 991.
85   Ray Desmond, *Dictionary of Irish and British Botanists and Horticulturalists* (London, 1994), p. 627.
86   Powell and Kennedy, 'Lord Kimberley and the foundation of Liberal Irish policy'.
87   See for example, John Devoy, *Michael Davitt: From the Gaelic American* (Dublin, 2008); Owen McGee, *The IRB: The Irish Republican Brotherhood: From the Land League to Sinn Féin* (Dublin, 2005).
88   P.J. Matthews, *Revival: The Abbey Theatre, Sinn Féin, the Gaelic League and the Co-operative Movement* (Notre Dame, Ind., and Cork, 2003); Sinéad Garrigan Mattar, *Primitivism, Science, and the Irish Revival* (Oxford, 2004); Gregory Castle, *Modernism and the Celtic Revival* (Cambridge, 2001).
89   Nicholas Allen, 'Review of Garrigan Matter', *Irish Studies Review*, 13 (2005), pp. 112–13.
90   For Lyons on Sigerson see endnote 1 above.
91   Lyons, 'Charcot's English translator', pp. 51–2.
92   Ibid., p. 50.

93   J.B. Lyons, *A Pride of Professors: The Professors of Medicine at the Royal College of Surgeons in Ireland* (Dublin, 1999), p. 99.
94   Christopher Lawrence, 'Incommunicable knowledge: Science, technology and the clinical art in Britain 1850–1914', *Journal of Contemporary History*, 20:4 (1985), pp. 503–20.
95   Lyons, *A Pride of Professors*, pp. 224–33. For a fuller account of Purser see also J.B. Lyons, *The Quality of Mercer's: The Story of Mercer's Hospital, 1734–1991* (Dublin, 1991).
96   Seamus Deane (ed.), 'Political writings and speeches 1850–1918' and 'Poetry 1890–1930'; both in Seamus Deane (ed.), *Field Day Anthology of Irish Writing* (3 vols, Derry, 1991), II, pp. 238–9, 281, 721, 728. For a further discussion of this debate see Luke Gibbons (ed.), 'Constructing the canon: Versions of national identity', in Deane (ed.), *Field Day Anthology*, III, pp. 950–5.
97   Luke Gibbons, 'Republicanism and radical memory: The O'Connors, O'Carolan and the United Irishmen', in Jim Smyth (ed.), *Revolution, Counter-Revolution and Union: Ireland in the 1790s* (Cambridge, 2001), pp. 211–37; idem., *Gaelic Gothic: Race, Colonization and Irish Culture* (Dublin, 2004).
98   Seamus Deane, *Strange Country: Modernity and Nationhood in Irish Writing since 1790* (Oxford, 1997).
99   David Dwan, 'Civic virtue in the modern world: The politics of Young Ireland', *Irish Political Studies*, 22:1 (March 2007), p. 56.
100 Ibid. For Young Ireland in this context see also Patrick Maume, 'Young Ireland, Arthur Griffith and republican ideology: The question of continuity', *Éire-Ireland*, 34:2 (Summer 1999), pp. 155–74; J. Quinn, 'John Mitchel and the rejection of the nineteenth century', *Éire-Ireland*, 38:3 (Fall/Winter 2003), pp. 90–108. For a wider contextualisation of nineteenth-century Irish republicanism see Margaret O'Callaghan, 'Reconsidering the republican tradition in nineteenth-century Ireland', in Iseult Honahan (ed.), *Republicanism in Ireland: Confronting Theories and Traditions* (Manchester and New York, 2008), pp. 31–42.
101 'Two groundbreaking new titles', www.Fielddaybooks.com, accessed 16/12/08, with reference to David Dwan, *The Great Community: Culture and Nationalism in Ireland* (Dublin, 2008).
102 Terry Eagleton, *Heathcliff and the Great Hunger: Studies in Irish Culture* (London and New York, 1995), p. 275.
103 Ibid., p. 274, n. 5, with reference to how in a particular scene in Bram Stoker's novel *Dracula* (1897), the 'bizarre blending of technology and Transylvania seems an apt symbol of the Irish mixture of tradition and modernity'.
104 Terry Eagleton, *Scholars and Rebels in Nineteenth-Century Ireland* (Oxford, 1999).
105 Ibid., p. 85.
106 Ibid., p. 86.
107 Luke Gibbons, '"Some hysterical hatred": History, hysteria and the Literary Revival', *Irish University Review* (Spring/Summer 1997), pp. 7–23.
108 Bram Stoker, *The Snake's Pass* (London, 1890).
109 For O'Halloran see Lyons, *Brief Lives*.
110 For Wilde see McGeachie, '"Normal" developments'.

111   For Wilde as Census Commissioner see Peter Frogatt, 'Sir William Wilde, 1815–1876: Demographer and Irish medical historian', in Eilean Ni Chuilleanain (ed.), *The Wilde Legacy* (Dublin, 2003), pp. 51–68.

112   For Madden see C.J. Woods, 'R.R. Madden, historian of the United Irishmen', in Thomas Bartlett, David Dickson, Daire Keogh and Kevin Whelan (eds), *1798: A Bicentenary Perspective* (Dublin, 2003), pp. 497–511.

113   For the English conversation see Leah Greenfield, *Nationalism: Five Roads to Modernity* (Cambridge, Mass., 1992).

114   For the debate on national styles of science see Maurice Crosland, 'History of science in a national context', *British Journal for the History of Science*, 10 (1997), pp. 95–113; Jonathan Harwood, 'National styles in science: Genetics in Germany and the United States between the world wars', *Isis*, 78 (1987), pp. 390–414; Roy Porter and Mikulas Teich, 'Introduction' to Porter and Teich (eds), *The Scientific Revolution in National Context* (Cambridge, 1992), pp. 1–10; Ludmilla Jordanova, 'Science and national identity', in Roger Chartier and Pietro Corsi (eds), *Sciences et langues en Europe* (Paris, 1996); idem., 'Science and nationhood: Cultures of imagined communities', in G. Cubitt (ed.), *Imagining Nations* (Manchester, 1998), pp. 192–211; Lewis Pyenson, 'An end to national science: The meaning and the extension of local knowledge', *History of Science*, 11 (2002), pp. 1–40.

115   Robert M. Young, 'Natural theology, Victorian periodicals and the fragmentation of a common context', in his *Darwin's Metaphor: Nature's Place in Victorian Culture* (Cambridge, 1985), pp. 126–64.

116   Paul White, *Thomas Huxley: Making the 'Man of Science'* (Cambridge, 2003). See also the discussion of White's book in Rebecca Scott, 'Masculinities in nineteenth-century science: Huxley, Darwin, Kingsley and the evolution of the scientist', *Studies in the History and Philosophy of the Biological and Bio-medical Sciences*, 35 (2004), pp. 199–207. For a more traditional interpretation of Huxley as the maker of professional scientific identity see Adrian Desmond, *Huxley: From the Devil's Disciple to Evolution's Priest* (London, 1998).

117   Lawrence, 'Incommunicable knowledge'.

118   For this earlier culture of care see Adrian Desmond, *The Politics of Evolution: Morphology, Medicine, and Reform in Radical London* (Chicago and London, 1989).

119   For scientific medicine see Bynum, *Science and the Practice of Medicine*.

120   L.S. Jacyna, *Medicine and Modernism: A Biography of Sir Henry Head* (London, 2008).

121   Ibid., pp. 3–4.

122   Castle, *Modernism and the Celtic Revival* and Garrigan Mattar, *Primitivism, Science, and the Irish Revival*.

7
'This Revived Old Plague'[1]: Coping with Flu

Caitríona Foley

Introduction

In nineteenth-century Ireland, influenza was among the staples of the GP's waiting room.[2] An outbreak in 1803, which 'overspread' Europe, was stated by the Local Government Board to have begun in Ireland, and the Board's report cited eight more visitations of the infection before 1850. The pandemic of 1847 also touched Ireland, compounding hardship in the most brutal year of the Famine.[3] Doctors noted that 'before this period the influenza was little more than known by name'. Its 'desperately overpowering influence had not hitherto been felt', but now it 'spread in a way which it had never been known to do before, and assumed a variety of shape and form quite new' to their experience; victims were 'calculated as more numerous than even those of the deadly cholera'. At a time when the 'full tide of death flowed on everywhere' in Ireland, however, outbreaks of flu seemed to draw little attention and seldom featured in the medical reports on the Famine.[4] Even some of the colloquial names given to flu, such as the 'old hin' and the 'homely malady', offer hints that influenza was not usually the kind of disease that evoked dread like fever, or that stirred panic like smallpox or cholera.[5]

In the late nineteenth and early twentieth centuries, however, Ireland was visited by pandemics of influenza which disrupted society, drove mortality sharply upwards, and left an extensive imprint on contemporary documents and reportage. The 'Russian flu' pandemic of the early 1890s has been described as 'the most severe influenza pandemic in the last three centuries', with the exception of the Great Flu of 1918–19.[6] A Ministry of Health report judged that the outbreak ushered in a 'new phase in the evolution of the disease', and January 1890, according to a Dublin doctor, was 'one of the sickliest ever experienced within living memory'.[7] When flu resurfaced in January 1892, the Irish death rate climbed to 19.4 per 1,000, the highest since registration had begun in 1864.[8] At the time, the Chairman of the Wexford Board of Guardians lamented that 'the mortality in the country is something dreadful', while the clerk noted that 'all the doctors were complaining, and the relieving officers had never before so many cases to attend to'.[9] The Great Flu of 1918–19 also unsettled perceptions of the malady; as *The Kerryman* put it, 'formerly when influenza happened along, it was looked upon as a kind of joke', but 'there is no fun about the present outbreak'.[10] Described by one historian as 'probably the greatest single natural disaster ever to hit this earth', the Great Flu was deemed to have 'far surpassed anything previously experienced', and left a residual dread of influenza among many who witnessed it first-hand.[11]

These outbreaks of influenza in the early 1890s and in 1918–19 brought the infection into sharp relief and offer an opportunity to explore the ways in which lay and medical communities responded to a disease that was often seen as part of everyday life.[12] This article explores the impact of these two epidemics, focusing on how the lay and medical communities coped with the disease, the treatments they used, and their understandings of infection and the sick body. It also examines the extent to which overlap or divergence was in evidence when it came to how the laity and medical professions conceptualised and dealt with disease. A final goal is to establish whether continuity or change were perceptible in popular and professional cultures of care over the period in question.

'An ancient visitor in a new garb': The Russian flu pandemic[13]

In the years leading up to 1890, influenza deaths in Ireland had been virtually negligible in terms of proportional mortality; in 1884–89, for instance, an average of only one death per 100,000 was attributable to the infection. However, when the Russian flu pandemic arrived in 1890, this rose to 36 deaths per 100,000, rising again in 1892, with 81 deaths per 100,000 due to flu. This was later eclipsed totally in 1918, when 10,651 flu deaths propelled the number to 243 per 100,000.[14] Striking hardest at the aged and the very young, the outbreaks of 1890–92 were consistent with the traditional social profile of the disease, as its casualties typically came primarily from the weakest groups in society.[15] In 1892, 5 per cent of flu mortalities came from those aged 55 and over, with deaths among children under ten years of age accounting for a further 17 per cent.[16] In the same year in Mitchelstown, county Cork, an estimated 60 per cent of the population was affected, with mortality 'fearfully high, but... altogether confined to aged people, and infant children'.[17] The master of Wexford Workhouse observed that 'when the old people are attacked they are not able to get over it', and Dr Leahy similarly pointed out to the Cork guardians that the flu casualties 'were, in most cases, old and debilitated'.[18] The 'young and the strong' were, 'as a rule, well able to do battle with it', although all classes seemed 'equally liable to attack', with 'princes and nobles' suffering alongside 'the humbler folk'.[19]

A report by a medical inspector of the Local Government Board in England traced this outbreak of flu to Bokhara, Russia, where flu had appeared in May 1889, following a 'succession of extreme meteorological conditions, and while the inhabitants were depressed in health owing to want of nourishment'. It further stated that an 'endemic centre' for influenza 'may exist in Russia, where, as we have seen, "la grippe" figures largely year by year as a cause of mortality'.[20] In his examination of the Russian influenza in the United Kingdom, F.B. Smith has noted that several other circumstances also indicated a Russian source for the flu. In 1889, low quality 'Russian oats' had been imported into London and the east coast ports. This grain was reputed to have caused sickness among the horses and domestic animals and fowl, with subsequent transmission of the illness to humans. In 1891, the epidemic was then judged to have been re-started by shiploads of Russian emigrants who landed at Hull for transhipment from Liverpool to New York, with flu spreading through Hull and into the east midlands. The introduction of the influenza to Liverpool in January via letters from sufferers in Russia offered further corroboration of the epidemic's eastern source.[21]

Its arrival in Ireland began to draw media attention in early 1890, with the *Irish Times* reporting a story that it had been introduced into the General Post Office in Dublin 'through the foreign mails'.[22] The Local Government Board report subsequently tracked the outbreak back to October 1889, when cases of flu were recognised 'in a few districts on the east coast of Ireland', spreading westwards over November. By January 1890 it 'prevailed generally throughout Ireland', peaking towards the end of the month and during the early part of February.[23]

Patients complained of sudden and biting chills, as if cold water was cascading down their backs and legs. Headaches, sleeplessness, occasional delirium, aversion to light, and loss of smell and taste constituted further symptoms. 'Extreme prostration, weakness, and nervous depression' were among the 'outstanding' features, with muscular pains, and pain in the eyeballs also prevalent. The mortality from influenza was judged by doctors to be due in 'large part' to a 'low and insidious form of pneumonia', which prevailed during the later phases of the epidemic.[24]

Large numbers sought help in hospitals and local dispensaries, and local medical officers were inundated with requests for attendance. The number of flu cases in 1892, which were 'severe enough to require relief by dispensary medical officers', reached 76,309. This figure, however, was estimated to have been 'but a fraction of the number which occurred', suggesting that the general population often treated themselves. Dublin hospitals experienced a significant increase in patient attendance at their dispensaries, and Jervis Street and Mercer's hospitals were 'especially tried in this respect', the former having 60 influenza patients one morning, although in neither institution 'were any considered of such severity as to necessitate detention'. Similarly at the Meath Hospital, although 'considerable call was made at the dispensary', no case was 'sufficiently serious' to be kept overnight, and the same could 'be said of the City of Dublin Hospital in Baggot Street'.[25]

The workhouse system came under stress in some areas as a result of the influx of flu sufferers. In the 1892 outbreak, the Mitchelstown workhouse hospital found itself with more than 'three times the ordinary number of sick persons', whilst the infirmary at New Ross, south Wexford, similarly became 'fearfully overcrowded' during the flu onslaught.[26] Entire families fell sick at the same time, sometimes receiving treatment at home from overwrought medical officers. In Camolin, county Wexford, 'three families were attacked with it, and they suffered very much, as they were all sick at the same time, and had no one to attend to them'. All over the county 'whole families' were laid up in January 1892, and 'the medical gentlemen of the districts' could 'scarcely cope with the numerous calls'.[27] In north Kerry, the doctors and the clergy were 'greatly overworked', especially in the Knockane, Duagh and the Gunsboro' districts, where the epidemic prevailed 'in its most malignant form'.[28] As the disease spread over Cork, the doctors at Schull were 'almost exhausted for want of proper rest'.[29]

When it came to understanding disease at this time, olfaction had a central role in both medical and lay conceptions of illness. Doctors believed that a 'noxious miasma' or 'abominable stench' could cause sickness, and folk tales of the Famine told of how people refused refuge to ill vagrants if they could 'smell the sickness from them'.[30] During the Russian flu pandemic, odours continued to be significant in both popular and professional medical understanding of sickness. Among medics, flu had been 'long regarded as of the "miasmatic group"', and there was a 'general consensus of opinion among medical officers' that the disease was 'airborne, and preceded and accompanied by high temperature and moist atmosphere'. Dr John Moore considered the 'virus of influenza' to be a 'miasma', or what 'sixteenth and seventeenth-century physicians' termed a 'fouling of the air'. January 1892 meanwhile saw a 'heavy, disagreeable fog...envelop' the Belfast area. This 'foul and choking vapour' was deemed by the Medical Superintendent Officer of Health in Belfast, Dr Whitaker, to be 'conducive to propagation of the influenza microbe'. This illustrates the way in which bacteriological conceptions of disease were grafted onto, rather than fully displacing, miasmatic interpretations. Although the principles of bacteriology were familiar to Irish doctors by the late nineteenth century, the medical profession was reluctant to completely discard the view that diseases had multiple causes. As was the case in Britain, bacteria and germs were meshed with the established medical 'contexts of airs, waters, and places'.[31] Similarly in France at this time, germ theory was fused with miasmatic ideas. A suggestion to use the Champ de Mars building as a temporary flu hospital was rejected out of a fear that 'if the floors of the galleries got permeated with influenza microbes', a 'chronic centre of infection would be created'.[32]

Another example of how bacteriology could be incorporated into miasmatic understandings of disease causation came from the pages of the *Belfast Evening Telegraph*. Here it was suggested that the reason the fishermen at Portstewart, Portrush, and along the northern seaboard had 'put out to sea at night in their usual health', but were 'unusually ill with influenza before morning', was that north-easterly winds had carried the disease from flu-stricken Galway. The paper dismissed the 'contact theory' of the disease, having

> little doubt that the specific germ of the disease propagates in such media as damp ground or air contaminated with organic exhalations, a confined watery atmosphere, such as that to which we must submit in dull, foggy weather.[33]

Consistent with medical ideas in Britain at the same time regarding the 'malarial atmospheric origins' of the disease, it was proposed in the

Irish Times that the 'gas set free from decayed matter' was to blame for the flu, the ailment having been

> prevalent during the last portions of our autumn when the fall and decay of all green matter has taken place, and the past few years we have had very wet autumns, particularly the month of December last...the extensive planting of evergreens round dwellings and open places would be of great benefit in absorbing the sickening vapours.[34]

John Roche, MD, retired from the Indian Medical Service, was similarly sure that the flu was 'due to the air', and was simply a 'species of malarial fever' typically 'prevalent after long periods of rain and floods, and also after long periods of drought in most countries'. Drawing on his first-hand experience with this type of infection – he had 'seen and suffered malarial fever' himself in the past – Roche further outlined how the rain had the effect of shutting up 'the surface of the earth', causing the air 'pent in beneath' to become 'foul' and injurious to health upon its release.[35] He advised that flu patients retire to bed and 'take sweating draughts and antiperrodics' to restore the body to health, subscribing to a humoral interpretation of disease, according to which patients could cure flu by adjusting the balance of fluids in their bodies.[36]

Odours were also considered important in terms of healing and defending the body against influenza. The *Mayo Examiner* described the popular belief that trees could 'suck up the malarious influence of marshes' that caused diseases like flu. An anonymous letter to the *Irish Times* declared the efficacy of a 'preparation of oil or eucalyptus', with proof of usefulness offered by the experience of a London insurance company 'who last year, supplied a number of clerks with this oil, to be ... dropped each morning, in their slices of blotting paper. Not one of the clerks took the ailment, while a number in another of their offices, who did not get the oil' contracted flu.[37] Advertisements for *Trench's Preventive Oil* highlighted its 'delicious' scent which also kept the 'Influenza Bacillus' at bay, while the makers of *Sanitas Oil* boasted that it could both prevent and cure influenza; all that the buyer had to do was to 'inhale and fumigate' with the product. A pharmacy in Rathgar, Dublin, also advertised its 'eucalyptus smelling salts for the prevention and cure of influenza'.[38]

The strengthen-and-stimulate thesis, characteristic of late nineteenth-century therapeutics in America, also informed lay and medical advice on how to cope with the infection. This emphasised the importance of diet in fortifying the body for its fight against disease, and of alcoholic beverages in invigorating the body's own natural defences.[39] The *Derry Standard* carried an advertisement which boasted that the humble *Bovril* was 'the best preventative and cure' for influenza.[40] Cod liver oil was also recommended in the pages of a Wexford broadsheet, along with whiskey punch and bed rest.[41] Doctors advised patients to 'take warm sustaining foods and drinks',[42] advice which was consistent with remedies used in earlier outbreaks of disease in Ireland, and with the medical and commercial cures recommended in Britain during the Russian flu outbreak.[43]

Alcohol meanwhile had been commonly used by both medics and the lay population to treat disease in the nineteenth century, and continued to be utilised during the epidemic.[44] 1892 saw a hefty increase in wine sales, with consumer demand guaranteeing sales of an extra 15,600 gallons of brandy.[45] Treatments employed by Dr John Moore in Dublin included 'free stimulation, frequent feeding, and quinine', while Surgeon P.H. Fox usually gave a

dose of calomel, or other purgative, to clear out the bowels, unless diarrhoea was present; the patient was then put on expectorants and milk or beef-tea diet; stimulants were given if the symptoms indicated their necessity, and quinine when the fever ran high; hot applications were applied to the throat and jacket poultices, turpentine, or spongio-piline to the chest.

John Daly & Co., Ltd., in Cork city advertised that its 'celebrated *Admiral Brand* Rum, if taken in time', would prevent the 'troublesome and sometimes dangerous' influenza from developing.[46]

There was also continuity with earlier centuries in lay perceptions of a link between spiritual and physical worlds. Religious modes of explaining illness endured during the epidemic, as faith, moral fortitude, and discipline were thought to offer a way towards health and healing.[47] The sickness experience was closely intertwined with religion, and a letter to the *Irish Times* in 1892 was in no doubt that the 'plague of sickness and death is but a working of Providence in the economy of the universe, and however good it may be to alleviate suffering, we should always keep in mind the Ruling Hand, and not vainly arrogate to ourselves powers we have not'. Other letters to the paper called for the bishops and clergy to hold special services and to offer 'special prayers to Almighty God in this time of sickness and mortality'.[48] Local newspapers advised their readers that 'there is nothing for those who are afflicted from the prevailing sickness to do but to trust to Providence'; 'men should trust in God's mercy a little more and not expect to have this world their own way'.[49] At a meeting of Dublin Corporation, Alderman Dennehy cautioned that 'they could not alter nor change the arrangements of Providence by showing this fear of catching disease'. Personal responsibility was vital, and Dennehy was convinced that 'the greater portion of those who died before him need not perhaps have preceded him had they taken ordinary care of their own health'. Temperance played its part in this regard: 'to preserve a man's health and energy to an advanced period he had only to live in accordance with the laws of God'.[50] This close association between spirit and matter was also part of traditional interpretations of disease outside of Ireland. Jonathon Barry for instance has found evidence that in eighteenth century Bristol, patients continued to seek 'a spiritual interpretation of illness'.[51] Beliefs regarding the spiritual dimension of sickness and healing were also characteristic of early nineteenth-century America, where people often viewed illness as an embodiment of God's displeasure and a warning to the dissolute.[52]

In Ireland, readers of *The People* (Wexford) were advised of the importance of living a temperate life, and avoidance of excess (in terms of alcohol consumption, over-exertion, over-eating etc.) was embraced as one of the best means of defence against influenza. The well-being of the body was popularly viewed as closely connected to lifestyle and as a system in balance, which could quickly break down if not treated properly.[53] A report on the pandemic commissioned by the British government similarly advised people on the need to avoid cold and fatigue, unwholesome food, and excess alcohol, and counselled that tiredness left one open to the infection.[54] Maintaining one's health demanded restraint and moral fortitude, while disease was a sign of individual transgression, indiscipline, and moral failure.[55]

The profusion of flu-specific pills and draughts in Irish newspapers, meanwhile, was consistent with the outburst of advertisements in British publications for influenza-related balsams, troches, and elixirs. It also indicated that the increasing commercial tendency towards self-care, evident in late nineteenth-century Britain, had spread to Ireland.[56] In the late nineteenth century, a boom in the sales of weekly and monthly magazines had taken place in the country, enabled by increasing literacy and prosperity, and an expanded rail network. Patent medicines featured frequently in the advertising space of these weekly and monthly publications, and also appeared in the pages of provincial newspapers, which numbered almost 220 in the latter half of the nineteenth century.[57] During the outbreak, newspapers such as the *Irish Times* and *Belfast Telegraph* carried advertisements for *Influenzicum* and *Pastille Bronchique* tablets, which were touted as the silver bullets that would prevent and treat the influenza. An 'absolute specific for this distressing malady' could be found apparently in 'M'Knight and Nicholl's Ammoniated Tincture of Quinine', available in Belfast pharmacies.[58] '*Trench*'s remedy for influenza and cold' involved putting a small drop of the liquid on the tongue every 15 minutes, which would cure the patient of the infection 'in a few hours'.[59]

The Russian flu outbreak of the early 1890s appears to have been marked by a considerable degree of overlap in terms of how the lay population and the medical profession dealt with influenza. Odours were viewed by lay and medical communities as important, both in terms of how disease spread and in relation to how it could be treated, while alcohol, food, and good living habits were also widely accepted as important in preventing and shaking off sickness. This body of ideas was also consistent with how disease was viewed and treated abroad in the same period. As Charles Rosenberg has noted, American physicians in the nineteenth century stressed the role of sustenance and alcohol in patient care, while the common use of patent medicines was also evident in Britain at this time. At the same time though, it should also be noted that, in Ireland, this shared language of disease existed alongside traditional interpretations of sickness. For instance, B.N. Hedderman, a district nurse on the Aran Islands off the west coast of Galway in the late nineteenth century, was vexed by the islanders continuing belief that the 'evil eye' or fairies could be responsible for illness, and that epidemics were simply 'proof of the wrath of the Almighty, or attributed to some magic power, or witchcraft'. Hedderman lamented that 'Pasteur himself would fail in teaching the "germ theory" here; their own therapeutic remedies are the preventives par excellence'.[60]

'The awakening': The Great Flu 1918–19

In the early weeks of the Great Flu, Dr Robert Watson warned people that 'the Russian influenza has been but a name to you. Even doctors have almost forgotten ... what influenza could do. And now comes the awakening'.[61] Reflecting on the 1918–19 pandemic in Great Britain and Ireland, the British Ministry of Health judged that 'we have just passed through one of the great sicknesses of history, a plague which within a few months, has destroyed more lives than were directly sacrificed in four years of destructive war'.[62] The scale of mortality was daunting as the outbreak claimed upwards of 50 million lives worldwide within the space of a year.[63] Plague was an oft-repeated trope of the pandemic, with newspapers blaring the word in their headlines. Rumours that the infection was a revived form of pneumonic plague were rampant, causing panics from Stockholm to Rio de Janeiro as the infection overwhelmed city districts and remote areas alike. In Montreal, in October 1918, there was such a surge in funerals that the cemetery officials could not 'dispose of the corpses fast enough, and at Mount Royal Cemetery at one time hundreds of dead lay there in their coffins unburied'. In Labrador, Newfoundland, owing to the 'amount of sickness among the survivors, the bodies of those who succumbed did not receive speedy burial, but were left lying about for a time; many of the corpses were, in consequence, devoured by dogs'.[64]

The pandemic washed over the world in three waves between spring 1918 and early summer 1919. The first wave arrived in Ireland in early June 1918, appearing among troops in Belfast and quickly spreading to surrounding munitions works and to the general population.[65] Reports of minor outbreaks in parts of Galway in the west soon followed, as did outbreaks in Dublin and Cork.[66] The disease remained largely confined to the east and north-east of the country and to more urbanised areas, with this summer outbreak lasting only a few weeks. Signs that a second wave was materialising were evident in Dublin from late September and, on this occasion, the infection spread rapidly over the rest of the country.[67] By Christmas 1918, all counties had experienced an outbreak, with both urban and rural areas witnessing the infection's gruelling effects. On the Inishowen peninsula, a sparsely populated region in north-west Donegal, one in four deaths in 1918 was due to influenza, while towns such as Naas and Newry suffered similar death rates.[68] Lingering in western counties until the end of 1918, this second wave of flu overlapped with a third wave which began to gather momentum at the end of January 1919, and lasted until April or May in the same year.[69]

Over the course of these three waves of influenza, the disease claimed upwards of 20,000 lives in Ireland, a figure marginally below the European average.[70] The Great Flu precipitated a level of sickness and death – 'so much havoc' – not witnessed in Ireland since the Famine years. Sean O'Casey described the piles of coffins outside undertakers' premises in Dublin at this time, 'towering barricades of them already sold, yet many more were needed for those who died'.[71] And in Ireland, as elsewhere around the globe, the virus struck hardest at those aged between 20 and 40 years of age, with 'the brunt of the 1918 pandemic ... borne by the young adult population'.[72] The strain of flu in circulation during the pandemic launched a much more violent assault on the bodies of its victims than had more recent strains of the infection. Doctors found themselves treating symptoms that had 'not been met with before in influenza', and in the *Lancet*, Dr William Collier commented that

> surely we are seeing a type of influenza quite different from any-thing we have seen before. I well remember the severe influenza epidemic of 1889–90, and attended a large number of cases, but the signs and symptoms which have been exhibited by patients I have attended during the past few days are quite new to me.[73]

A 'pronounced feature' of the serious cases was the heliotrope cyanosis, the way in which the body turned a bluish-purple due to the presence of oxygen-deficient blood. People whispered of the rapid decompos-ition of flu victims and the necessity of hasty burials, as 'in a few cases the faces of the victims became discoloured some time after dying'.[74] Many patients 'bled from the mouth and ears', and nose bleeds were a common early symptom.[75] Some complained of pains 'down the front of the chest behind the sternum as though (they) were "all raw inside there"', and coughed up frothy, blood-stained or purulent sputum. Autopsies revealed lungs to have been reduced to 'gruyere cheese' con-sistency, resembling the morbid changes endured by war-time victims of gassing.[76]

At this time in Ireland, the central health authority was the Local Government Board. Its most visible response to the outbreak was to make 'acute influenzal pneumonia' a notifiable disease in spring 1919, by which time the third wave was well underway.[77] The lack of respon-siveness to the epidemic was not unique to Ireland though; as Wilfred Witte notes, when it came to responding to the outbreak, few govern-ments could claim to have achieved anything 'significant worth men-tioning in this regard'.[78] In Dublin, Dr Kathleen Lynn lamented the sluggishness of official reactions to the epidemic, despairing of the Local Government Board and of the Dublin Corporation.[79] Elsewhere around the country, the reaction of local authorities, the boards of guardians, and urban and rural district councils, varied from vigorous to indifferent, with civil society providing the most effective response to the epidemic.[80] Relief committees sprang up in numerous towns around Ireland, often driven by the 'ladies' of the community.[81] Decentralisation was also characteristic of the response by the nationalist organisation Sinn Féin to the Great Flu, with relief activity taking place at grass roots level with various Volunteer and Cumann na mBan branches providing aid to the sick.[82]

At this time, administration of medical relief in Ireland was based on the dispensary system. Under this system, the country was divided into 723 dispensary districts, each with one or more medical officers (doctors) who were answerable to the poor law guardians. The doctor was obliged to attend every sick person in the district in possession of a dispensary ticket, which was provided for each episode of sickness by a poor law guardian. There were two kinds of ticket: a black one entitled the holder (or his dependants) to attend the doctor at the dispensary; while a red ticket, also known as a 'scarlet runner', meant that the holder could receive medical attendance at home.[83] Although hospitals and workhouses overflowed with patients during the epidemic, the home appears to have been the natural locus of care of the sick, with home attendance by medical officers almost doubling during the year of the epidemic. There was a documented increase of more than 100,000 home cases treated by dispensary doctors, a figure which may not even have been representative of the true case numbers, since Local Government Board officials guessed that some doctors had been too over-run to fully record their case-loads.[84]

In Tipperary during the outbreak, home-care was the norm. Michael MacCarthaigh, for instance, remembered that 'patients remained in their homes', and that the local doctor was 'night and day on the road in his back-to-back car (horse-drawn of course)'.[85] There were also examples of a reluctance among rural inhabitants to be brought to hospital: the ambulance driver for Boyle Union reported to the guardians that he had made two or three journeys for some patients in the Arigna mountains, 'and they did not come with him'.[86] The medical officer of Dripsey district in Cork reported that 'nearly all the cases have refused to go to hospital', even though several were 'very bad' and had to be seen by him twice a day. He eventually 'got a few cases to consent to go to Hospital'.[87] The high death rate among flu victims at Cork Street Fever Hospital, where almost one in four flu cases died, suggests that many may have been brought there simply as a last resort.[88] The Great Flu thus indicates that popular fears of the hospital which had frustrated nineteenth-century doctors and nurses, continued well into the twentieth century.[89]

At the same time, however, the authority of, and reliance on, professional medical aid was evident from instances such as that in Manorhamilton, county Leitrim, where 'there was a regular panic before Dr Reynolds came to do duty'. In Claremorris, county Mayo, the local doctor fell ill, with the result that one of the officials had people coming to him 'every other day looking for a doctor'.[90] A member of Clonakilty Board of Guardians observed that during the epidemic 'there was a row of cars every morning' at the doctor's door. In parts of the country, the epidemic may have even served to enhance the authority of medical officers, with doctors earning the gratitude of the local people and authorities. A member of the Galway Union Board of Guardians explained at a meeting that he had been 'directed by some of the people in Turloughmore district to propose a resolution thanking Dr Glynn for his work during the terrible epidemic in that district'. The chairman supported the resolution, asserting that 'the epidemic was very severe in the district', and that he had heard 'some of the people say that were it not for Dr Glynn they would not know what to do'.[91] The epidemic also provided evidence, however, that families were not afraid to question doctors' treatment, as several complaints were made to boards of guardians that doctors had been negligent or inattentive. This reflects perhaps the ambivalence with which medical officers were viewed in this period, as observed by George Bernard Shaw:

> the noble character of the profession and the honour and conscience of its members, ... the tragedy of illness at present is that it delivers you helplessly into the hands of a profession you deeply mistrust.

Although the prestige of doctors was increasing, the 'standard of sobriety' was still 'below what it ought to be', and doctors were also noted for their attendance at races and the local hunt.[92]

In terms of the social response to the epidemic, the repeated waves of infection elicited much fear, with the result that in some areas community bonds were eroded and the usual voluntary offers of care and help were not forthcoming. In Schull Union, west Cork, the clerk of the Board of Guardians reported a flu case that was a particularly 'bad one, and nobody would go near the house'. Around Skibbereen, when awareness spread 'that some unfortunate person or family had got the influenza ... their houses were at once isolated ... nobody would come near to offer them assistance or comfort of any kind'.[93] In both urban and rural parts of Wexford, 'many of the homes of the poor which have been visited by the influenza scourge' were shunned; 'everyone' appeared 'to be afraid of catching the contagion'.[94] The medical officer of Wicklow dispensary district reported to the guardians that 'the large number of the community that were stricken down and also the fear of contagion prevented us from obtaining unskilled help of any kind'. One of the nurses in Rathdrum Union similarly reported that 'no attendant would come near to work in the wards' as they were 'afraid of the disease'.[95]

At the same time, there were many instances of philanthropy, as communities pulled together to help care for the worst hit sections of society, especially the poor and the isolated.[96] The *Connacht Tribune* reported on the harsh reality of the epidemic for those living on Lettermullen and Gorumna on the Galway coast:

Their needs are few and humble; yet even these needs cannot be served. Practically all on the islands became afflicted with the influenza. Poorly fed, many succumbed. No spring work has been done ... who is to sow the potatoes, who is to pay the rent or to feed the people for the coming year?[97]

A doctor from Tuam, county Galway meanwhile told the local Board of Guardians that they had

no conception of the state of affairs that existed. I went into one house at half past three in the morning, and found a woman lying dying on the kitchen floor. Two children were staggering about hardly able to stand, and two more children were dead. By the aid of my electric hand-light I made my way into the room, and there I found a man lying dead across a bed. There was no light or anything in the house. You have no conception how [sic] things were. You got the nurse to help these people, and it was a good thing. The epidemic was an exceptional one. Even if you tried the most you could recover would be a few shillings from those patients.[98]

Some of the most harrowing examples of the effects of the flu came from Longford, where the death rate from the disease was 'very great' and the 'conditions of distress ... appalling'. In response, the Bishop called a meeting of the parishioners to 'relieve the alarming condition of affairs, and a respectable sum was promptly subscribed and a committee appointed to relieve the distress'. One of the guardians noted that

the position ... in Longford was the gravest ever I remember ... the relieving officer was absolutely up to his eyes in work ... and it was humanly impossible for him to deal with the great quantity of distress that was existing in the town.

The Relieving Officer reported that house-to-house visits had to be organised 'simply because people were not able to go out of their houses to make applications' for relief. The dead bells were 'ringing all the day in Longford', with people 'dying without either the doctor or the relieving officer to know they were dying'. The doctor himself admitted to the guardians that he was 'not able to attend, and did not know, all the sick cases in Longford', with another guardian confirming that 'the doctors themselves are knocked over by it'. The chairman spoke of a general anxiety in evidence: 'this thing is very infectious and they are shaking going into houses'.[99]

The idea that the environment had a direct role in spreading the flu stimulated a rush to clean up public spaces, with large-scale disinfection and clearing operations launched in several towns under the Public Health (Ireland) Act of 1878.[100] The Act had established sanitary districts for the purposes of public health, and gave local authorities the power to deal with a wide variety of 'nuisances' arising on public health and other grounds. Examples of typical nuisances included offending pools, ditches and drains. The Act also empowered authorities to destroy unsound food, supervise slaughter houses, and isolate persons suffering from certain infectious diseases.[101]

The measures taken by local authorities, as well as the advice provided by doctors, suggest the endurance of sanitarian beliefs among both the laity and the medical profession during the Great Flu. Stressing the part played by an unclean environment and adverse social conditions in disease propagation, the sanitarian creed had dominated medical thought and public health theories in mid-nineteenth-century Europe. Sanitarians advised that disease could thus be prevented by ensuring clean surroundings, an absence of 'nuisances', and efficient waste disposal, advice which was echoed during the Great Flu.[102] During the second wave of flu, Dr Moorhead, the medical officer of Tullamore district, stressed that people should pay 'attention to personal and household cleanliness, ventilation and limewashing, scavenging of lanes and streets'. Superintendent Medical Officer for Dublin, Sir Charles Cameron, advised the flushing of sewers, lavatories and yards with Jeyes' fluid and carbolic acid.[103] These measures were consistent with precautions taken in Spain and Italy in the same period, where large scale use of disinfectant, street sweeping, and refuse disposal were carried out in order to curb the spread and effects of flu.[104]

Sanitarian ideas were also folded into older conceptions of disease relating to sickening air and odours. A doctor based in Cloyne, county Cork, considered that the origins of the epidemic lay in 'fomites emanating from decaying and dead animal matter'. These were 'the progenitors of the most active of the fever-generating germs, and identical with the contagion found in the foetid putrescent atmospheres of houses, rooms, and places were filth and squalor abound'.[105] In Belfast, the street and ashpit cleaners went on strike during the second wave of the outbreak, and animal and vegetable refuse could be seen 'piled up waist high' in the streets, 'creating an atmosphere conducive to disease'. The danger 'from the fermenting accumulations' was feared to be growing daily, with councillors calling attention to the matter in the press.[106]

There was also continuity in the enduring importance of odours when it came to treating the disease, with eucalyptus in particular enjoying widespread popularity, cast in the same protective role as it had been during the Russian flu outbreak. Henry Foran and Dermot J. Stewart, the Joint Secretaries of the Operative Tailors, Irish Tailors and Tailoresses Union, Dublin, wrote a letter to the *Freeman's Journal*, decrying the employment by 'many master tailors and other employers in the clothing trade' of 'home workers' to produce garments, mentioning how

> you see in the streets, trams, public conveyances, etc., members of the public with eucalyptus-laden handkerchiefs held close up to the face to prevent infection from the air we breathe, but quite impervious to the fact that the garments in which they are closely muffled may be impregnated with millions of disease microbes, owing to the conditions and places in which they were produced.

In Enniscorthy, 'every second person one meets with reeks of eucalyptus', and even Charles Cameron counselled people to 'keep a little pad of eucalyptus and smell it often'.[107] Other healing and preventive scents included 'creosote or coal-tar exposed in a saucer, or flat receptacle, in a sick room give off fumes which, inhaled by the patient, relieve a cough and prevent the spread of infection'. The Wexford health committee advised that 'peppermint essence' was, alongside eucalyptus, 'a useful precaution'.[108]

The epidemic also suggested that assimilation of the principles of bacteriology, the scientific study of micro-organisms, was continuing to take place among doctors and laity alike. For instance, a depot was opened at 37 Charlemont Street, Dublin, for 'Inoculation against Influenza, under the stewardship of Dr Kathleen Lynn'.[109] Vaccines were also used in a preventative capacity by medical officers around the country, reflecting the increasing importance of the laboratory in the medical profession, and providing evidence that the initial reluctance to take up germ theory among doctors was giving way to acceptance by 1918.[110]

At this stage, the vocabulary of bacteriology was also filtering through more extensively into popular discourse. In a letter to the *Freeman's Journal*, the traditional association of disease with the lower classes surfaced once more as the working classes were scapegoated for spreading the flu, due to their practice of '"blowing the top" off a frothy tankard of stout', which was thought to be one of the ways in which the infection was being spread:

> Beer and stout germinate rapidly, and when a bar-tender who may be recovering from an attack of the prevailing epidemic breathes his germ-laden breath on a glass of stout the consequences are too awful to contemplate. The deadly microbes are, so to speak, sent direct from 'grower to consumer'. It is quite possible this dreadful practice is responsible for the appalling death rate amongst the working class.[111]

The passage reflected the way in which the path of disease had often functioned as a 'legitimisation of the social order', with infection traditionally seen as originating among the lower classes, and subsequently contaminating the higher social strata.[112] This was also underlined by anxieties about the morals and conduct of the working classes, echoing the theories of cleanliness from the eighteenth and nineteenth centuries, which had been distinguished by their ideological function, striving to maintain not just health standards but the social order as well.[113]

When it came to the commercial response to the epidemic, cures and preventatives for flu were often promoted on the basis of their bacteriological and germ-killing properties.[114] The relatively widespread demand for disinfectants during the epidemic (Ballymote, county Sligo, reportedly almost ran out of supplies) further reflected the ongoing incorporation of germ theory and the principles of bacteriology into popular understandings of disease.[115] Medical advice on how to avoid flu also overlapped with fears and directives about tuberculosis. Sir Charles Cameron stressed that sputum contained flu microbes, and spitting in public places in Dublin was banned for a time during the outbreak.[116] Other medics emphasised the need to keep utensils clean, and to wash them separately from those of the flu patient, further reminiscent of measures advocated in Ireland and America to prevent spread of TB. Yet another similarity between measures taken in response to the two diseases was evident when the circulation of library books was stopped in Dublin in 1918; their pages could act as carriers of flu germs in the same way as they functioned as fomites of TB.[117]

The importance of individual responsibility was another theme that informed medical and official perspectives on treatment, and further indicated absorption of germ theory ideas into popular conceptions of health. The *Roscommon Herald*, for instance, suggested that

> something should be done to compel persons in whose houses the disease is prevalent to remain at home – if they have not the charity and decency to do so themselves – until the danger has passed, and not be the means of spreading the disease, as may be the case when they mix with others in shop, church, hall, or train.[118]

Dr Fleury, in an interview with the *King's County Chronicle*, emphasised that 'a great deal depended on the individual himself ... if the sufferer took to bed at once, and received proper medical treatment and attention, he would in all probability come through the attack safely'.[119]

The close links between health, disease and religion also continued, as communal prayers were offered around the country in the hope that the disease would abate.[120] A concerned citizen from Monasterevan wrote to the *Freeman's Journal* suggesting a 'general wearing of the Sacred Heart badge' as a way of fending off the flu.[121] Faith in the healing power of God and the providential aspect of illness was even articulated at the meetings of local government bodies: the chairman of Castlebar urban council spoke to his fellow council members about his trust 'in God that they would be spared any further deaths as a result of this dread disease', and the Council collectively expressed its hope that 'it would please Providence to stay the ravages of this terrible epidemic'. Obituaries of flu victims similarly referred to 'Providence the Great Healer of all suffering'.[122]

The outbreak disrupted church services in some areas, and the Archbishop of Dublin advised the clergy that, 'as long as the present danger of infection continues, the solemn service of the church should not be prolonged by the chanting of the Office of the Dead'. He also suspended fast and abstinence on the Vigil of the Feast of All Saints, in view of the 'depressed condition of the public health'. Similarly the Bishop of Waterford and Lismore dispensed with fasting and abstinence owing to the epidemic, and in Tuam, county Galway, fasting obligations were suspended during Lent due to the arrival of the third influenza wave.[123]

Food and drink continued to be viewed as important influences on health and convalescence at this time. An advertising campaign for *Bovril* recalled the Russian flu outbreak, as the beef stock was promoted on the basis of its supposed ability to fortify the body against influenza. A letter to the *Freeman's Journal* meanwhile stated that 'the cure for influenza' was a combination of milk, food, and heat.[124] A local paper in Offaly referred to the common belief that 'the strange epidemic of influenza attacked only those who were run down because of the lack of proper food'. It went on to note, however, that this belief had been 'exploded' when a dispatch in August to Washington from an Irish port

> told of the occurrence of symptoms of this disease among officers and men stationed at an American destroyer base. Apart from American soldiers, the American sailors are probably the best fed persons in Europe, but the disease attacked several score of them there, and for a week or so disrupted crew assignments completely.[125]

Garlic, meanwhile, was a popular folk remedy. In Sligo, faith in the plant's ability to fend off the illness was 'still so widespread at the time of the 1918 influenza pandemic that many carried a piece around with them in a pocket'. On the Aran Islands, it could be eaten, placed under the pillow, or even carried in one's clothing to keep flu away.[126]

In Ireland and abroad, alcohol was used widely to combat the Great Flu. It was the most popular remedy in Britain, while in Spain alcoholic liquors such as brandy or 'cazalla' were the sole nourishments given to flu victims.[127] Dr Sexton recommended to the Cork guardians that workhouse inmates employed at the mortuary 'get meat, eggs, butter and two pints of porter daily, during the epidemic of influenza',[128] and elsewhere in Cork, 'whiskey proved, if not an effective, at least a popular and palatable medicine'.[129] D.W. Macnamara, who was a junior doctor in the Mater during the outbreak, reflected that whiskey or brandy in 'heroic doses' had been a particularly popular option among 'the older men'. Linda Kearns, a district nurse in Achill in the epidemic, lost no patients to the flu, and attributed her success to her 'use of poitín as medicine'.[130]

Flu directives in newspapers and from doctors commonly cast illness in qualitative terms, as being due to the cold or wet, and drew on ideas of the body as a system of intake and outgo. *The People* (Wexford) advised people not to 'get wet feet', and 'if you do, walk home if possible to avoid packed conveyances. Wear warm loose clothing'.[131] Doctors advised that the body could expel the flu infection by using emetics or by inducing perspiration. A Cork physician found that 'a hot diffusible stimulant combined with good intestinal evacuant' was 'infinitely better than drugs', and considered 'copious drinking of hot nourishing liquids, such as whey, barley-water, and milk, hot water with a little milk added, and fowl broth' to be advisable. Dr Robert Watson stressed that it was 'important to get the bowels well-opened at the outset and regularly throughout the course of the disease', and 'the more briskly the patient can sweat the better for him'; this was one of the surest ways to eliminate the 'poisons of the disease' from the system.[132] Emetics were possibly already in use by the laity, as doctors had noted during the Famine that they were 'at all times a favourite remedy with the country folk'.[133]

The reality was, however, that avoiding exposure to the disease in the first place was the surest way to prevent flu. Given the 'inability of official medicine to provide specific treatment for the disease', the best advice that doctors could give was to stay in bed, as nursing gave patients 'the best possible chance to survive'. With the nature of the disease still not fully understood, the most effective treatment for flu at this time was simply care and rest.[134]

Conclusion

An examination of the extensive outbreaks of influenza in the late nineteenth and early twentieth centuries suggests that the period was marked by considerable continuity in terms of how both doctors and 'ordinary' people dealt with the flu-stricken body. This is consistent with Charles Rosenberg's finding in relation to late nineteenth-century American doctors, who sought to ensure the 'greatest possible degree of continuity with older ideas' in both the intellectual realm and in practice.[135] In both the Russian Flu and the Great Flu, the home was the usual site of treatment, although workhouses and hospitals witnessed spikes in attendance during outbreaks. Taking care of oneself in the first place, by getting sufficient rest, eating healthily, practising good personal hygiene, and living morally, was viewed as the best way of ensuring that one remained safe from a flu attack. A healthy, clean, fit body was the most effective protection against flu in both medical and popular views, with discipline and self-regulation championed as intrinsic to health maintenance. The condition of the body also had a spiritual dimension, as prayers were offered on behalf of the sick and in the hope of persuading God to help expunge epidemics. Another way of improving one's defences was to make sure that the immediate environment was clean and unpolluted, with the belief that infection coincided with 'physical squalor and moral degradation' being evident during both outbreaks.[136] The profusion of advertisements for flu cures during both outbreaks, meanwhile, reflected the increasing commodification of health in this period. Change was evident, albeit to a lesser extent, in the shape of increased acceptance and assimilation of germ theory in the medical and lay responses, as suggested by doctors' willingness to use, and people's willingness to receive, vaccines in 1918–19, as well as by the increasing use of the vocabulary of bacteriology in everyday language.

An examination of medical and popular discourse regarding influenza in the late nineteenth and early twentieth centuries suggests that when it came to beliefs regarding infection and disease, doctors and laymen made use of similar languages in their efforts to read sickness and to make the human body legible. 'Elite notions' of disease and health were not dramatically removed from the pool of knowledge called on by ordinary men and women in their attempts to understand sickness and to protect and heal their bodies. The ideas, interpretations, and theories stimulated by influenza outbreaks indicate that there were several 'points of cross-over' in popular and elite medical cultures when it came to coping with a 'revived old plague'.[137]

Notes

1 *Irish Times*, 15 January 1890.
2 Tony Farmar, *Patients, Potions and Physicians: A Social History of Medicine in Ireland 1654–2004* (Dublin, 2004), p. 112.
3 *Eighteenth Annual Report of the Local Government Board for Ireland 1890* [C. 6094], H.C. 1890, xxxiv, p. 22; *The Irish Builder*, 15 October 1897, p. 183.
4 'Report upon the recent epidemic fever in Ireland', *Dublin Quarterly Journal of Medical Science*, 8 (1849), pp. 310–11, 270; see also 'Report upon the recent epidemic fever in Ireland', *Dublin Quarterly Journal of Medical Science* (hereafter *DQJMS*), 7 (1849).
5 J.M. Synge, *The Aran Islands* (London, 1907/1992), p. 16; *Irish Times*, 4 January 1890; *Copies of the Correspondence between the Local Government Board and its Inspector with regard to the Present Epidemic of Smallpox at Athenry* [Cmd. 422], H.C. 1875, lx; S.J. Connolly, 'The "blessed turf": Cholera and popular panic in Ireland, June 1832', *Irish Historical Studies*, 23:91 (1983).
6 John M. Barry, *The Great Influenza* (London, 2004), p. 261.
7 Ministry of Health, *Report on the Influenza Pandemic 1918–19* (London, 1920), p. v; John William Moore, 'The influenza epidemic of 1889–90, as observed in Dublin', *Dublin Journal of Medical Science*, 89:220, 3rd Series (April 1890), p. 301.
8 F.B. Smith, 'The Russian influenza in the United Kingdom, 1889–1894', *Social History of Medicine*, 8:1 (1995), p. 64.
9 *The People* (Wexford), 10 February 1892, 3 March 1892.

10   Ministry of Health, *Report on the Influenza Pandemic*, p. xviii; *The Kerryman*, 26 October 1918; 'During the last few years influenza was in one of its non-virulent periods, and was treated lightly. Now the old virulence has shown itself once again', *Down Recorder*, 2 November 1918.

11   Colin Brown, 'The influenza pandemic of 1918 in Indonesia', in Norman G. Owen (ed.), *Death and Disease in Southeast Asia* (Oxford, 1987), p. 235; Ministry of Health, *Report on the Influenza Pandemic*, p. vi; Patrick Logan, *Making the Cure* (Dublin, 1972), p. 27.

12   See for instance B.N. Hedderman, *Glimpses of My Life in Aran: Some Experiences of a District Nurse in these Remote Islands, off the West Coast of Ireland* (Bristol, 1917), p. 82.

13   *Munster Express*, 30 January 1892.

14   Ministry of Health, *Report on the Influenza Pandemic*, p. 52.

15   Barry, *The Great Influenza*, p. 238; in Ireland, influenza had a close association with older age groups, and the Census Commissioners of 1841 noted that it was mentioned in ancient manuscripts under the name *Creatan*, deriving from *Creat*, the thorax, 'in which latter sense it is understood to be incidental to old age'.

15   Sir William J. Thompson, 'Mortality from influenza', *Journal of the Statistical and Social Inquiry Society of Ireland*, 14 (1920), p. 5.

16   *Twenty-ninth Annual Report of the Registrar-General* (Ireland), [C. 7255], H.C. 1893–4, xxi, pp. 8, 14.

17   *Cork Examiner*, 19 January 1892.

18   *The People* (Wexford), 10 February 1892; *Cork Examiner*, 8 January 1892.

19   *The People* (Wexford), 16 January 1892; *Cork Examiner*, 8 January 1892.

20   Dr Parsons, *Report on the Influenza Epidemic of 1889–90* [C. 6387], H.C. 1890–91, xxxiv, p. 102.

21   Smith, 'The Russian influenza in the United Kingdom', p. 66.

22   *Irish Times*, 2 January 1890, 3 January 1890, 4 January 1890.

23   *Eighteenth Annual Report of the Local Government Board for Ireland* [C. 6094], H.C. 1890, xxxiv, p. 24.

24   Moore, 'The influenza epidemic of 1889–90, as observed in Dublin', pp. 310, 312, 315; Ministry of Health, *Report on the Influenza Pandemic*, p. vii.

25   Parsons, *Report on the Influenza Epidemic*, p. 12; *Irish Times*, 9 January 1890.

26   *Cork Examiner*, 19 January 1892; *The People* (Wexford), 3 February 1892.

27   *The People* (Wexford), 2 January 1892, 9 January 1892, 16 January 1892.

28   *Kerry Sentinel*, 6 February 1892.

29   *Cork Examiner*, 9 January 1892.

30   'Report upon the recent epidemic fever in Ireland', *DQJMS*, 8, p. 279, and 7 (1849), pp. 86, 111; Cathal Poirteir, *Famine Echoes* (Dublin, 1995), p. 112; Deborah Brunton, 'Dealing with disease in populations: Public health, 1830–1880', in Deborah Brunton (ed.), *Medicine Transformed: Health, Disease and Society in Europe, 1800–1930* (Manchester, 2004), pp. 190, 199.

31   Parsons, *Report on the Influenza Epidemic*, pp. x, 12; Moore, 'The influenza epidemic of 1889–90', p. 313; *Belfast Evening Telegraph*, 19 January 1892; Farmar, *Patients, Potions and Physicians*, p. 101; Smith, 'The Russian influenza pandemic', p. 62.

32 *Irish Times*, 22 January 1892.
33 *Belfast Evening Telegraph*, 18 February 1892.
34 Smith, 'The Russian influenza in the United Kingdom', p. 62; *Irish Times*, 12 January 1892.
35 *Irish Times*, 15 February 1892.
36 Michael Worboys, 'Colonial and imperial medicine', in Brunton (ed.), *Medicine Transformed*, p. 219.
37 *Mayo Examiner*, 23 January 1892; *Irish Times*, 20 January 1892.
38 *Belfast Evening Telegraph*, 2 February 1892; *Irish Times*, 9 February 1892.
39 Charles Rosenberg, *Explaining Epidemics* (Cambridge, 1992), p. 26.
40 *Meath Herald*, 9 January 1892; *Derry Standard*, 27 January 1892.
41 *The People* (Wexford), 9 January 1892.
42 *Irish Times*, 6 January 1890.
43 Poirteir, *Famine Echoes*, p. 108; Cathal Poirteir, *Glortha on nGorta* (Dublin, 1996), p. 138; Roger J. McHugh, 'The Famine in Irish oral tradition', in R. Dudley Edwards and T. Desmond Williams (eds), *The Great Famine: Studies in Irish History, 1845–52* (Dublin, 1956), p. 416; Lori Loeb, 'Beating the flu: Orthodox and commercial responses to influenza in Britain, 1889–1919', *Social History of Medicine*, 18:2 (2005), p. 208.
44 McHugh, 'The Famine in Irish oral tradition', p. 417; *Report of the Commissioners of Health, Ireland, on the Epidemics of 1846 to 1850* [Cmd. 1562], H.C. 1852–3, xli, p. 73.
45 *Irish Times*, 22 February 1892.
46 Moore, 'The influenza of 1889–90, as observed in Dublin', p. 311; P.H. Fox, 'Report on influenza and brief abstract of cases treated at the Station Hospital, Arbour Hill, Dublin', *Dublin Journal of Medical Science*, 90 (1890), p. 43; *Cork Examiner*, 15 January 1892.
47 Roy Porter, 'Introduction', in Roy Porter (ed.), *Patients and Practitioners: Lay Perceptions of Medicine in Pre-Industrial Society* (Cambridge, 2002), p. 7.
48 *Irish Times*, 12 January 1892, 19 January 1892, 20 January 1892.
49 *The People* (Wexford), 16 January 1892; *Mayo Examiner*, 6 January 1892.
50 *Irish Times*, 7 January 1890.
51 Jonathan Barry, 'Piety and the patient: Medicine and religion in eighteenth-century Bristol', in Porter (ed.), *Patients and Practitioners*, p. 172.
52 Paul Starr, *The Social Transformation of American Medicine* (New York, 1982), p. 36.
53 *The People* (Wexford), 9 January 1892; similar advice was offered in the *Derry Standard*, 25 January 1892.
54 *Belfast Evening Telegraph*, 18 February 1892.
55 Rosenberg, *Explaining Epidemics*, p. 114.
56 Smith, 'The Russian influenza in the United Kingdom', p. 69; Margaret Pelling et al., 'The era of public health, 1848 to 1918', in Charles Webster (ed.), *Caring for Health: History and Diversity* (Buckingham, 2001), p. 102.
57 Farmar, *Patients, Potions and Physicians*, p. 125; Marie Louise Legg, *Newspapers and Nationalism: The Irish Provincial Press, 1850–1922* (Dublin, 1999), p. 13.
58 *Irish Times*, 14 January 1890; *Belfast Evening Telegraph*, 2 January 1892.
59 *Irish Times*, 22 January 1892.
60 Hedderman, *Glimpses of My Life in Aran*, pp. 84, 88, 80.
61 *Ulster Guardian*, 13 July 1918.

62   Ministry of Health, *Report on the Influenza Pandemic*, p. xviii.
63   Barry, *The Great Influenza*, p. 450.
64   Ministry of Health, *Report on the Influenza Pandemic*, pp. 208, 338, 275, 280.
65   *Belfast Telegraph*, 11 June 1918; *The Lancet*, 13 July 1918, p. 51.
66   *Western News*, 22 June 1918; *Sligo Champion*, 29 June 1918; Cork Poor Law Union, Board of Guardians Minute Books, BG 69 A 149, pp. 315, 321, Cork City and County Archives.
67   Royal Irish Constabulary (hereafter RIC) Reports, CO 904/107, p. 311 (University College Dublin); *Irish Times*, 10 October 1918.
68   *Fifty-fifth Detailed Annual Report of the Registrar-General for Ireland* [Cmd. 450], H.C. 1919, x, pp. 52, 54.
69   *Tipperary Star*, 11 January 1919; *The Kerryman*, 18 January 1919; *Mayo News*, 18 January 1919; *Fermanagh Times*, 23 January 1919; *The People* (Wexford), 22 January 1919.
70   Niall P.A.S. Johnson and Juergen Mueller, 'Updating the accounts: Global mortality of the 1918–1920 "Spanish" influenza pandemic', *Bulletin of the History of Medicine*, 76 (2002), p. 113.
71   William J. Thompson, 'Mortality from influenza in Ireland', *Journal of the Statistical and Social Inquiry Society of Ireland*, 14 (1920), p. 1; Peter Somerville-Large, *Irish Voices* (London, 1999), p. 20.
72   Ministry of Health, *Report on the Influenza Pandemic*, pp. 52, 90–1.
73   *Ulster Guardian*, 23 November 1918.
74   Ministry of Health, *Report on the Influenza Pandemic*, pp. 66, ix; *Freeman's Journal*, 28 October 1918; *The People* (Wexford), 2 November 1918.
75   *The People* (Wexford), 2 November 1918.
76   Ministry of Health, *Report on the Influenza Pandemic*, pp. 72, 100–1.
77   CSO RP/1919/7278, National Archives of Ireland (NAI); in contrast, the notification of influenza was made obligatory in New York State as early as 11 October 1918, coming into force on 14 October 1918, Ministry of Health, *Report on the Influenza Pandemic*, p. 301.
78   Wilfred Witte, 'The plague that was not allowed to happen', in Howard Phillips and David Killingray (eds), *The Spanish Influenza Pandemic of 1918: New Perspectives* (London, 2003), p. 57.
79   Folder 6, Dr Lynn's papers (106/59), p. 4 Allen Library, Dublin.
80   In the town of Macroom, county Cork, for instance, 'very laudable and desirable measures' were taken by the Urban Council in order to 'mitigate the suffering and distress consequent upon the ravages of the influenza epidemic', *Cork Examiner*, 8 November 1918. By contrast, in a wire to the Westport Board of Guardians, county Mayo, a parish priest chastised the local authorities for their inaction, *Mayo News*, 1 March 1919.
81   See for instance *Freeman's Journal*, 8 November 1918, 11 November 1918; *Meath Chronicle*, 23 November 1918; *The People* (Wexford), 2 November 1918.
82   Lil Conlon, *Cumann na mBan and the Women of Ireland* (Kilkenny, 1969), p. 70; Michael O'Suilleabhain, *Where Mountainy Men Have Sown* (Tralee, 1965), p. 48; *Freeman's Journal*, 5 November 1918.
83   Ruth Barrington, *Health, Medicine and Politics in Ireland 1900–1970* (Dublin, 1987), pp. 8–9.

84 *Annual Report of the Local Government Board for Ireland 1918–19* [Cmd. 578], H.C. 1920, xxi, p. 28.
85 Michael MacCarthaigh, *A Tipperary Parish* (Midleton, 1986), p. 176.
86 *Roscommon Herald*, 28 December 1918.
87 Cork Poor Law Union Board of Guardians Minute Books, BG 69 A 150, pp. 81, 111, Cork City and County Archives.
88 *Sixty-first Annual Report of the Board of Superintendence of the Dublin Hospitals* [Cmd. 480], H.C. 1919, xiii.
89 McHugh, 'The Famine in Irish oral tradition', p. 415; 'Report on the recent epidemic fever in Ireland', *DQJMS*, 7, pp. 374–6; Hedderman, *Glimpses of My Life in Aran*, pp. 82–3.
90 *Fermanagh Times*, 20 March 1919; *Mayo News*, 9 November 1918, p. 2.
91 *Cork County Eagle*, 22 March 1919; *Connacht Tribune*, 25 May 1919.
92 *The People* (Wexford), 13 February 1919; *King's County Chronicle*, 28 November 1918; Farmar, *Patients, Potions and Physicians*, p. 125; Barrington, *Health, Medicine and Politics*, pp. 10–11; Caitriona Clear, *Social Change and Everyday Life in Ireland, 1850–1922* (Manchester, 2007), pp. 92–3.
93 *Cork County Eagle*, 21 December 1918, 16 November 1918.
94 *The People* (Wexford), 16 November 1918, 23 November 1918, 30 November 1918.
95 Minute Books of Rathdrum Union 1918, PLUR/WR/136, p. 438; Minute Books of Rathdrum Union 1919, PLUR/WR/137, p. 126, Wicklow County Archives.
96 *Down Recorder*, 29 March 1919; *The People* (Wexford), 2 November 1918.
97 *Connacht Tribune*, 29 March 1919.
98 *Connacht Tribune*, 26 April 1919.
99 *Roscommon Herald*, 30 November 1918.
100 *Mayo News*, 14 December 1918 and 9 November 1918; *King's County Chronicle*, 14 November 1918; *The People* (Wexford), 25 January 1919; Minutes of Rathdrum Rural District Council, WLAA/RDCR/M/18 (May 1918–December 1918), p. 420, Wicklow County Archives.
101 Barrington, *Health, Medicine and Politics*, p. 12; http://www.enfo.ie/leaflets/fs12-6.htm (24 January 2009).
102 Margaret Pelling, Mark Harrison, and Paul Weindling, 'The Industrial Revolution, 1750 to 1848', in Webster (ed.), *Caring for Health*, p. 75; 'nuisances' designated offences against the environment which were injurious to neighbours or to the community at large, Margaret Pelling and Mark Harrison, 'Pre-industrial healthcare, 1500 to 1750', in Webster (ed.), *Caring for Health*, p. 42.
103 *King's County Chronicle*, 19 December 1918; *Freeman's Journal*, 29 October 1918.
104 E. Tognotti, 'Scientific triumphalism and learning from facts: Bacteriology and the "Spanish flu" challenge of 1918', *Social History of Medicine*, 16:1 (April 2003), p. 102.
105 Mary E. Fissell, 'Making meaning from the margins', in Frank Huisman and John Harley Warner (eds), *Locating Medical History: The Stories and their Meanings* (Baltimore, 2004), p. 376; *Mayo News*, 14 December 1918.
106 *Freeman's Journal*, 2 November 1918.

107   *Freeman's Journal*, 5 March 1919; *The People* (Wexford), 30 November 1918; *Freeman's Journal*, 20 February 1919.
108   *The People* (Wexford), 30 October 1918.
109   *Freeman's Journal*, 2 November 1918.
110   *Donegal Vindicator*, 24 January 1919; Terrie M. Romano, 'The cattle plague of 1865 and the reception of "the germ theory" in mid-Victorian Britain', *Journal of the History of Medicine and the Allied Sciences*, 52 (1997), pp. 53–4; Farmar, *Patients, Potions and Physicians*, pp. 101–2.
111   Deborah Lupton, *The Imperative of Health* (London, 1997), p. 37; *Freeman's Journal*, 4 November 1918.
112   Pierre Bourdieu, *In Other Words: Essays towards a Reflexive Sociology* (Oxford, 1990), p. 135; Thomas Willis, *Facts Connected with the Social and Sanitary Condition of the Working Classes in the City of Dublin* (Dublin, 1845), pp. 50–1.
113   Lupton, *The Imperative of Health*, pp. 34–6.
114   *Freeman's Journal*, 16 July 1918, 30 October 1918, 15 November 1918, 18 February 1919.
115   *Roscommon Herald*, 2 November 1918; *The People* (Wexford), 2 November 1918; Paul Weindling, 'From germ theory to social medicine', in Brunton (ed.), *Medicine Transformed*, p. 245.
116   Katherine Ott, *Fevered Lives. Tuberculosis in American Culture since 1870* (London, 1996), pp. 112, 119; Greta Jones, 'The campaign against tuberculosis, 1899–1914', in Elizabeth Malcolm and Greta Jones (eds), *Medicine, Disease and the State in Ireland, 1650–1940* (Cork, 1999), pp. 160–1.
117   Ott, *Fevered Lives*, pp. 81, 116; Weindling, 'From germ theory to social medicine', p. 254.
118   Pelling et al., 'The era of public health, 1848–1918', p. 100; *Roscommon Herald*, 30 November 1918; see also *King's County Chronicle*, 24 October 1918. Ministry of Health, *Report on the Pandemic*, p. 196.
119   *King's County Chronicle*, 24 October 1918.
120   *Mayo News*, 21 December 1918; *King's County Chronicle*, 7 November 1918; *Freeman's Journal*, 26 October 1918, 28 October 1918.
121   *Freeman's Journal*, 30 October 1918.
122   *Mayo News*, 16 November 1918, 23 November 1918.
123   *Derry People and Donegal News*, 9 November 1918; *Irish Independent*, 4 November 1918; *Church of Ireland Gazette*, 15 November 1918, p. 765; *Irish Catholic*, 2 November 1918; *Cork Examiner*, 30 October 1918; *Connacht Tribune*, 29 March 1919.
124   *Freeman's Journal*, 5 March 1919.
125   *King's County Chronicle*, 19 December 1918.
126   David E. Allen and Gabrielle Hatfield, *Medicinal Plants in Folk Tradition* (Cambridge, 2004), p. 328; many thanks to Dr Susan Kelly for bringing this reference to my attention. Ruairi O hEithir, 'Folk medical beliefs and practices in the Aran Islands, Co. Galway' (MA thesis: University College Dublin, 1983), pp. 106, 131.
127   N.P.A.S. Johnson, 'The overshadowed killer: Influenza in Britain in 1918–19', in Phillips and Killingray (eds), *The Spanish Influenza Pandemic of 1918*, p. 152; Beatriz Echeverri, 'Spanish influenza as seen from Spain', in Phillips and Killingray (eds), *The Spanish Influenza Pandemic of 1918*, p. 180.

128 Cork Poor Law Union Board of Guardians Minute Books, BG 69 A 150, p. 484, Cork City and County Archives.
129 O'Suilleabhain, *Where Mountainy Men have Sown*, p. 48; John Kavanagh, 'The influenza epidemic of 1918 in Wicklow town and district', *Wicklow Historical Society Journal*, 1:3 (1990), pp. 36, 40.
130 D.W. Macnamara, 'Memories of 1918 and "the 'Flu"', *Journal of the Irish Medical Association*, 35:208 (1954), p. 306; Proinnsios O Duigneain, *Linda Kearns: A Revolutionary Irish Woman* (Leitrim, 2002), p. 23.
131 Andrew Wear, 'Puritan perceptions of illness in seventeenth-century England', in Porter (ed.), *Patients and Practitioners*, p. 83; Rosenberg, *Explaining Epidemics*, p. 13; Brunton, 'Introduction', p. xi; *The People* (Wexford), 22 February 1919; for similar advice, see *The Kerryman*, 16 November 1918.
132 *Mayo News*, 14 December 1918, p. 3; *Ulster Guardian*, 13 July 1918, p. 4.
133 'Report upon the recent epidemic fever in Ireland', *DQJMS*, 7, p. 85.
134 Tognotti, 'Scientific triumphalism', p. 106; Barry, *The Great Influenza*, pp. 319, 359.
135 Rosenberg, *Explaining Epidemics*, p. 28.
136 Nikolas Rose, 'Medicine, history and the present', in Colin Jones and Roy Porter (eds), *Reassessing Foucault* (London, 1999), p. 56.
137 *Irish Times*, 15 January 1890. David Armstrong, *Political Anatomy of the Body* (Cambridge, 1983), pp. 1–2; Jonathon Barry, 'Piety and the patient: Medicine and religion in eighteenth-century Bristol', in Porter (ed.), *Patients and Practitioners*, p. 145; Porter, 'Introduction', p. 19.

8

'Half mad at the time': Unmarried Mothers and Infanticide in Ireland, 1922–1950[1]

Clíona Rattigan

Introduction

Mary Bridget B., a 22 year old unmarried woman, was tried for the murder of an unnamed male infant at the Central Criminal Court in October 1944.[2] She gave birth alone in her bedroom in June 1944. Mary Bridget's sister and mother only discovered that she had given birth the following day when Mary Bridget asked her sister to send for a doctor 'because she was feeling weak and wanted a doctor as she had a baby'.[3] When Dr Anne Fitzgerald arrived at the family home on 18 June 1944, Mary Bridget's mother 'shut the door and said "this is an awful case Doctor. We have a very bad case here." [Dr Fitzgerald] asked what was wrong with the girl. The mother said "Oh now it's a bad case for you." [She] asked "is it a confinement case by any chance?" The mother replied "it's a confinement case and an awful one."'[4]

This chapter will look at the role played by medical professionals in infanticide cases tried at the Central Criminal Court, Dublin, between 1922 and 1950 by focusing on the depositional evidence of doctors and midwives who treated single women suspected of killing their infants. I use the term 'infanticide' where appropriate throughout this chapter, even though infanticide legislation was only passed in the 26 counties in 1949. This chapter examines infanticide cases tried in the first few decades of post-independence Ireland, when both church and state were deeply concerned about Irish morality. The importance of doctors in the witness stand in infanticide cases in Ireland was well established by the twentieth century. Indeed their statements, in relation to the mental health of unmarried women who had given birth unassisted, may have affected the verdict and the sentence in a number of cases. This chapter also examines encounters between unmarried infanticidal women and medical professionals. The ways in which female defendants and doctors interacted highlight the power of the medical profession and the vulnerability of poor women suspected of murdering their illegitimate infants. This chapter will also consider the criminal activity of an Irish midwife who ran a private nursing home in Dublin during the 1930s. As Quinn has observed, in relation to English infanticide cases during the late Victorian period, there are a 'wide variety of other voices present in these narratives'.[5] This study is not limited to the depositions of medical or legal professionals and, by drawing on the statements of unmarried women who were charged with infanticide, and the depositional evidence submitted by their relatives and employers, it is possible to build a more detailed picture of the physical and emotional trauma experienced by unmarried women who gave birth in Ireland during the years 1922 to 1950.

My research is based on the records of four cases heard at the Court of Criminal Appeal and the records of 191 cases tried at the Central Criminal Court between 1922 and 1950. Verdict and sentencing information for defendants in this sample was sourced in Trials Record Books for the period under consideration. Many more cases were heard in the lower courts during the period under review. Arnot and Usborne have observed that 'historical studies in different contexts on topics such as childcare, infanticide and abortion have shown that there could be major disjunctures between official penal prescription and popular perceptions and practices'.[6] That there was a disjuncture between church teaching and the law on the one hand, and the attitudes held by many Irish people with regard to illegitimate child life on the other, is evident from my examination of infanticide cases in the Irish Free State. In many instances, both the single, infanticidal woman and her relatives displayed a complete lack of regard for the lives of children born outside wedlock. Some do not seem to have considered the murder of illegitimate infants a serious crime. Presumably some single women in this sample felt that taking their infants' lives was preferable to bringing up an illegitimate child on their own, given the stigma associated with single motherhood in post-independence Ireland. Interestingly, some women, as in the case of Mary M., were concerned about baptising their infants immediately before taking their lives. Mary M. told Superintendent Sean Murphy that she did not want anyone to know anything about the birth of her illegitimate infant. She explained how she had baptised her infant 'in the name of the father and of the son and of the holy ghost before putting it in a bucket of water'. She then 'said the apostles' creed and threw holy water on it'.[7] According to the Catholic Church, 'the newborn child has an inalienable right to life, quite as much as a grown boy or girl, man or woman, and no difference can be recognised in the murder of the one or the other'.[8] Writing in February 1929, the Catholic Bishop of Ossory stated that 'these principles should be known to our people and our legislators'.[9] He recognised the fact that, in Ireland at that time, the illegitimate child was 'less valued than the life of the ordinary child' but he stated that 'this is not law or morality'.[10]

Infanticide cases often came to light in Ireland when a woman who chose to give birth alone and in secret subsequently required medical attention and had to send for a doctor, as in the case of Mary Bridget B. Doctors were often the first people to come into contact with unmarried mothers who had committed infanticide. When they treated these women in their bedrooms, they were often the first to enter what had, in effect, become a crime scene. Moreover, doctors were also often the first to attempt to find the baby's body. As soon as she reached Mary Bridget's bedroom, Dr Fitzgerald 'took down the bedclothes that were over the accused in bed and [she] said [she] wanted to find the baby'.[11] Mary Bridget's mother then pointed to a suitcase on the bedroom floor. When Dr Fitzgerald opened the suitcase she 'found a blood-stained piece of brown paper covering a male child's body'.[12] Doctors were also often the first to report a suspected case of infanticide to the Gardaí. Midwives too feature in the records of Irish infanticide trials. In a number of infanticide cases, a nurse, rather than a doctor, was the first person to treat an unmarried mother suspected of infanticide.

Medical certificates

Doctors clearly played an important role in infanticide cases tried in Ireland during the first half of the twentieth century and were regularly called upon during the initial stages of investigations into suspected cases of infanticide, as they could easily establish whether or not a woman had recently given birth. In a 1933 county Limerick case an infant's body was found in the grounds of a mental hospital. Bridget M. was questioned two days later. When Sergeant Peter Higgins asked her if 'she had any objection to be examined by a medical doctor', she said 'no'.[13] In fact, Bridget's mother advised the Gardaí to 'take her to a doctor and have finished with it'.[14] Following the medical examination Sergeant Higgins charged Bridget with murder and concealment of birth.

Many women denied that they had given birth when they were first questioned. The records of infanticide cases indicate that Gardaí may have attempted to procure confessions, by threatening to take women suspected of committing infanticide to a doctor in order to determine whether they were telling the truth. In a 1934 county Monaghan infanticide case Bridget C., the accused, claimed that the guard who took her statement intimidated her and threatened her.[15] According to Bridget, when she denied that she had given birth, the guard responded by saying, 'I will soon make you tell, I will bring you to see the doctor'.[16] It was unclear whether Bridget C.'s statement had been a voluntary one. When the case was tried at the Central Criminal Court the judge criticised the guard involved and, in particular, the 'method in which the Sergeant took her evidence'.[17] He also remarked that 'some of the Guards think that it is only necessary to rattle off the caution in a glib way'.[18] Many of the women in this sample may well have been unaware of their legal rights.

County homes

Most unmarried women charged with the murder of their new-born infants during the period under review gave birth alone in their parents' home or their employers' home, and medical professionals who treated infanticidal women usually encountered them in their bedrooms in their parents' or employers' homes. However, a minority of unmarried women in this sample gave birth in public hospitals. Fewer still gave birth in private nursing homes. As Earner-Byrne has noted 'the prevailing view of unmarried motherhood' in Ireland between 1900 and 1950 'was that it was illegitimate, unsustainable and morally wrong'.[19] Indeed, the language of crime was used extensively to describe unmarried mothers, as Kennedy has noted: 'the language used about unmarried mothers during the first half of the twentieth century in Ireland was marked by the colours of crime and sin'.[20] Kennedy highlights that phrases such as 'rehabilitation of the mother' and 'girls who have fallen again' were used repeatedly in government reports.[21] O'Sullivan and Raftery have also noted that 'the language of criminality was deliberately invoked' in discussions of unmarried mothers 'highlighting the perception of such women as "criminal"'.[22] When unmarried mothers were mentioned in the Irish press during the first half of the twentieth century, they were often characterised in negative terms and referred to as 'troublesome',[23] or as 'offenders'.[24] Articles on single mothers in the Irish press often carried headings such as, 'A difficult problem. The Unmarried Mother Question',[25] or 'Unmarried Mother Problem'.[26] Single Irish women who entered a county home or union workhouse to give birth generally lied about their marital status and registered as married women, mobilising one of the strategies they could use to deflect unwanted attention by medical staff and officials in the county homes and hospitals.

From the evidence given in the Elizabeth E. and Christina R. cases, it would appear that unmarried mothers were treated differently from married women with new-born children by staff in maternity hospitals. Unmarried expectant women were sometimes refused treatment altogether, as in the case of Ellen D. When she was tried for infanticide in November 1930, Ellen D. said that she had been refused admission to the county home in her area. In reaching a verdict of not guilty, the foreman of the jury explained that 'the jury felt that the unfortunate girl took steps to be attended properly at her confinement, but was not successful'.[27] He stated that refusing admission to a pregnant woman was deplorable and suggested that 'steps should be taken to have the matter remedied'.[28] Unmarried mothers immediately aroused the suspicion of staff in county homes, hospitals and unions. In some instances, members of staff in these institutions seem to have been hostile towards single, pregnant women. Christina R. managed to avoid detection for the duration of her stay in the South Dublin Union in August and September 1930. The ward mistress who was questioned in court explained that staff at the South Dublin Union did not pay as much attention to married mothers as to unmarried mother. 'You see this woman was supposed to be a married woman and we did not take as much notice of her as if she had been a single woman.'[29]

Unmarried mothers, by contrast, were observed closely. Members of staff carefully monitored the ways in which they interacted with their new-born infants. When called on to give evidence during Christina R.'s trial, Mrs Margaret O'Callaghan detailed a conversation she had with Christina in her house on Capel Street the day she left the Dublin Union. 'I then asked her if she was married and she said no. I asked her how she got out of the union as I understood unmarried girls would have to stay some time in the union to mind their babies.'[30] It would appear that unmarried mothers were considered a high-risk category by staff in the Dublin Union and in county homes nationwide. It was assumed that unmarried mothers were far more likely to abandon, neglect or harm their children than married mothers. Judging by the statistics on infanticide in this period, these fears were not without foundation. The vast majority of women who stood trial in the Central Criminal Court for the murder of their infants between 1922 and 1950 were unmarried. In a report to the Carrigan committee, Hannah Clarke, an NSPCC inspector, claimed that the Gardaí were 'now keeping an eye on girls in trouble to prevent infanticide'.[31] She stated that a sergeant in Maynooth 'had ten cases under observation recently'.[32]

An article in the *Irish Times* on 27 March 1935 described how staff at the Roscommon County Home had repeatedly warned Elizabeth E. against neglecting her daughter. 'A nurse in that institution had warned the mother not to abandon it or the guards would be after her', and 'Dr C.J. Kelly, Roscommon, said that he warned Elizabeth E. that if she did not take care of the child she would get into trouble'.[33] Staff in Ireland's network of county homes may have warned unmarried mothers about taking care of their children but women like Elizabeth E. were not offered any concrete support or given much useful advice. In a 1939 county Tipperary case, Dr Mary Courtney, who attended to Mary B., a 23 year old unmarried mother, recalled how she had tried to help Mary.[34] Shortly after Mary's baby was born, Dr Courtney made enquiries regarding the adoption of her infant and managed to find three addresses for her. However, in most cases there is no evidence to suggest that doctors or nurses offered any practical advice. In Dublin and other large cities, unmarried mothers could turn to societies who helped women to foster their children. However, it was far more difficult for unmarried mothers in rural areas with tight knit communities to adopt that kind of solution.

Nurses, like doctors, were involved at various stages of investigations into suspected cases of infanticide. In a 1938 Dublin case an identification parade was held at the Bridewell Garda station. Two nurses were asked to identify a woman they had treated some weeks previously. Both nurses identified Mary M., who had been a patient in St Kevin's hospital and in St Patrick's hospital. According to the nurses, Mary M. had had a baby boy with her at the time. Nurses who treated unmarried mothers charged with infanticide were often asked to make statements. In some cases the nurses who admitted unmarried mothers to hospital, or discovered the bodies of their infants, appear to have been quite concerned about the women in their care. Nurse Deborah Stack attended to Joan C. the morning after she gave birth and, in her statement, she expressed a strong sense of concern for Joan's physical and mental health. Nurse Stack said that Joan was in 'a weak condition suffering from shock' and that she 'was in a dazed condition and did not seem to realize anything'.[35] For the most part, however, the depositional evidence of nurses was factual. Their statements were usually brief and there was often little to suggest that nurses sympathised with the plight of unmarried mothers or tried to advise or assist them.

Nurses like Mary Anne Cadden and Florence Lang, who ran their own private nursing homes, inadvertently provided an important service for unmarried expectant women, who hoped to keep their pregnancies a secret.[36] Unmarried expectant women who were able to pay the fees charged in private nursing homes chose to give birth there because the homes were small and quiet and it was unlikely that neighbours, employers or friends would discover that they had given birth. Mary Kate M. went to such a nursing home in Galway to have her baby in March 1943. In his statement, John M., the father of the child, described the plans they made. 'Before she left home it was our intention that she should give birth to the child at some place where it would not become known that the child had been born'.[37]

The Gardaí who questioned Mary Anne Cadden about the abandonment of an infant in county Meath in June 1938, implied that most of the patients in her nursing home were unmarried mothers. In her statement, Cadden maintained that she had not admitted an unmarried mother since 1936, although she added that 'there may have been unmarried girls here and bring men with them and say they are married but I would not know'.[38] However, two unmarried women, who had given birth in her nursing home in March 1938, were later questioned. It is clear from their depositions that they had been extremely anxious to avoid publicity and scandal. Annie H. explained that '[her] idea in coming to Dublin was to cover up the shame of it for the sake of [her] family'.[39] Given the negative attitudes towards unmarried motherhood in Irish society during the years 1922 to 1950, it is hardly surprising that unmarried mothers would have sought privacy and anonymity. Nursing homes like the one run by Mary Anne Cadden in Rathmines provided a way for unmarried mothers to avoid further humiliation and shame.

Following an investigation in July 1938, it emerged that a number of unmarried mothers who gave birth in nurse Cadden's nursing home had paid her 50 pounds to have their new-born infants adopted by a community of nuns. However, Cadden may have abandoned the infants. She was charged with obtaining money by false pretences and unlawfully abandoning two infants. Bridget M., the aunt of Annie H., an unmarried mother who gave birth in Cadden's nursing home, told Gardaí that '[she] believed that the child was taken to a Catholic home. [She] would not have paid the fifty pounds unless [she] believed that the defendant would have carried out her promise'.[40] Patrick C. covered the cost of his sister's stay in Cadden's nursing home and, when she was admitted, he spoke to Cadden about adopting the baby. He recalled how 'she next said do you wish to have the baby adopted and that if so the mother would have to sign away any further interest in the baby for a sum of fifty pounds'.[41] Patrick C. understood that nuns would place the baby with a couple who did not have any children. At the time '[he] thought [his] sister was in excellent hands and [he] returned home satisfied'.[42] Nurse Cadden seems to have exploited the vulnerability and desperation of unmarried mothers and their relatives. In June 1938 Cadden and her employee Mary O'Grady were charged with conspiring to abandon a child and with abandoning a child. They were tried at the Central Criminal Court in May 1939. Nurse Cadden was sentenced to 12 months imprisonment with hard labour. A report in the *Irish Times* noted that nurse Cadden's 'means of livelihood in Dublin was gone forever' and that she 'was to all practical purposes ruined'.[43] In July 1939 the Central Board of Midwives removed Mary Anne Cadden from the roll of midwives. The records from Cadden's trial provide a rare glimpse into the manner in which single women from wealthier backgrounds dealt with an unplanned pregnancy in this period. Private nursing homes, like the one run by Cadden in Dublin, enabled women from wealthier backgrounds to give birth without being interrogated about their marital status in the way that poorer women were questioned when they entered county homes to give birth. Elizabeth E. wore a wedding ring when she was admitted to the Roscommon County Home and she gave a false name. However, she was repeatedly quizzed by a midwife about her name and eventually had to admit that she was single.

In a 1933 county Clare case, the professional conduct of a nurse was called into question. In July 1933 Norah M., an unmarried woman gave birth in her parents' home. Nurse Mary M. was called to the house just before Norah gave birth and she was present when the infant died. Mary M. was suspected of involvement in the infant's death and she was charged, along with Norah M. senior and Norah M. junior, with murder, conspiracy to murder, concealment of birth, and conspiracy to conceal the birth of Norah's unnamed female infant. At first nurse Mary M. denied having been called to attend to Norah M. but she later admitted to having been present when the girl gave birth. Mary M. also admitted that she had helped bury the infant. Dr Denis O'Dwyer said that 'there was a strong suspicion of suffocation' in the case but nurse Mary M. was eventually cleared of all charges and was discharged on 28 November 1933.[44] Norah M. senior was sentenced to serve two months in prison.

It is not possible to determine nurse Mary M.'s exact level of involvement in the case. She may simply have been a witness to a crime committed in Norah M.'s home but, although Mary M. pronounced the infant dead, she chose not to report the infant's death. However, Mary M. did advise Norah M. senior to send her daughter to the workhouse but Norah M. senior refused to do so because she did not want her husband to find out about her daughter's pregnancy. It is possible that nurse Mary M. may have taken pity on Nora M. and her daughter and helped them to conceal the birth of Norah's baby. In doing so she risked serving a prison sentence and ruining her own career.

Cases involving nurses like Mary M. and Mary Anne Cadden undermine traditional notions of cultures of care and point to the possibility that some nurses, like Mary M., may have been involved in conspiracies to conceal the birth of the infants of unmarried mothers. Unscrupulous nurses like Mary Anne Cadden were willing to profit from unmarried mothers. However, the actions of nurses like Mary M. and Mary Anne Cadden seem to have been the exception, rather than the rule, in infanticide cases during the period under review.

Infanticide and class

It is clear from the records of infanticide cases tried at the Central Criminal Court between 1922 and 1950 that a number of women charged with infanticide had difficulty expressing themselves. Most single women charged with infanticide or concealment of birth during this period were from working class backgrounds and were employed as domestic servants, while a number of women came from extremely disadvantaged backgrounds. Many of these women had received very little, if any, formal education and some were completely illiterate. Coupled with the physical and psychological strain of giving birth alone, it is hardly surprising that the statements of unmarried mothers charged with the murder of their new-born infants, or with concealment of birth, can appear confused. Arrests were generally made shortly after the women involved had given birth, which, in most cases, had been without medical assistance and followed months of concealing their pregnancies from family, friends and employers. The doctor who examined Mary S., a 22 year old unmarried woman who gave birth unassisted in her employer's home, said that when he saw her 'she was in a very confused condition and not really mentally capable of making any reliable statement'.[45]

Doctors who treated single women suspected of committing infanticide often commented on the level of intelligence of their patients. Unmarried mothers who had concealed the birth of their infants were sometimes perceived as stupid, ignorant or mentally defective. In a report written on 14 June 1929 a medical officer stated that Deborah S. was 'of low mental calibre'.[46] The medical officer who examined Mary M. in Mountjoy prison in September 1948 reported that he 'found her childish in manner' and that, while she had attended school until she was 13 years old, 'she shows a very low standard of education and is unable to do simple calculations'.[47] She never read newspapers or books and took no interest in general affairs. Mary was also poorly informed about pregnancy and childbirth and the medical officer concluded that she was both mentally immature and mentally defective.[48] Another Mary M., a 21 year old domestic servant, was tried at the Central Criminal Court for the murder of her unnamed female infant in April 1941. Dr James Maher spoke to Mary when she was admitted to the Coombe hospital. In his deposition he noted that Mary was not inclined to speak at the time. He also added that 'from her attitude to the questions I thought she was a bit dull mentally'.[49] This patronising attitude by well-educated, affluent doctors seems to have been rooted in class bias. Insensitive remarks made by doctors reveal a lack of sympathy and understanding, and suggest that some may have been ill-equipped to deal with women who may have committed infanticide.

Two single women charged with the murder or concealment of birth of their illegitimate infants between 1922 and 1950, Edith A. and Margaret F., were both considered to be mentally defective. The medical officer who examined Margaret F. in Sligo prison concluded that she was a mental defective and she was found insane at her trial in December 1943. Edith A., according to the medical officers who examined her in Mountjoy prison, was mentally defective and they came to the conclusion that she was unfit to plead. In a letter to prison authorities, the rector in Edith's parish stated that she had always been mentally defective and 'though she went to school regularly for a good many years could never learn anything and is almost or quite unable to read or write'.[50] In his deposition Edith's father said that he did not think she was capable of taking care of herself. He also stated that she had never been "normal". Edith's stepmother was of the opinion that Edith was not accountable for what she did. She also maintained that Edith was not normal. However, comments were made about the level of intelligence of a number of other women in this sample. The doctor who examined Mary M. said that 'in [his] opinion the girl is barely normal'.[51] In his report on Mary M. the prison medical officer noted that while she was fit to plead 'the mentality of this prisoner is low'.[52] Dr Thomas McDonagh said he thought that Jane M. was 'sub-normal in intelligence'.[53] The medical officer who examined Frances B. in Mountjoy prison noted that she was 'dull' and 'slow in replying to questions'.[54]

The law and infanticide

As well as confirming that a woman suspected of murdering her new-born infant had recently given birth and determining the cause of death, doctors also gave evidence on the suspects's mental state. In a 1944 county Kilkenny case, Dr Valentine Coughlan stated that Bridget D.'s infant 'died of asphyxia due to a cord being tied around its neck' but when cross-examined Dr Coughlan agreed with the barrister defending Bridget D., who argued that it is 'a well known fact that immediately after child birth the mother's nervous and mental condition is frequently abnormal'.[55] Dr Coughlan told the court that this condition could last for several days. While many doctors shared this opinion that childbirth disturbed a woman's nervous and mental condition, until 1949 the deliberate killing of a fully delivered infant was considered no different from murder in legal terms and carried a mandatory death penalty under Section 1 of the 1861 Offences Against the Person Act. However, only a small number of cases prosecuted in the Central Criminal Court during the period under review actually resulted in murder convictions. The death sentence passed on women convicted of the murder of their infants between 1922 and 1949 was always commuted, and most women convicted of this crime spent only a relatively short time in prison. The passing of the Infanticide Bill in 1949 did not mark a significant change in the way in which the courts dealt with women convicted of the murder of their infants. This legal change did, however, spare the indicted woman, the judge, the jury and the government 'all the distress and worry of an unreal trial for murder followed by reprieve' and sanctioned existing practice.[56] As McAvoy has observed, 'it is clear that the introduction of infanticide legislation was not perceived as a priority issue by any Irish government prior to 1949'.[57] She has suggested that this low priority was due to the fact that contemporary legal practice had ensured that most infant murder prosecutions ended in lesser verdicts.

Between 1922 and 1949, 12 women were sentenced to death for the murder of an illegitimate infant. Seven were unmarried mothers. Three of the women sentenced to death were female relatives of the birth mother. One woman was married and one woman was a widow. In all 12 cases, the convicted women were recommended to mercy and their sentences were reduced to penal servitude for life but few served more than three years in prison. Of the seven unmarried mothers who were sentenced to death between 1922 and 1949, five were recommended to mercy because of the 'exceptional', 'overwhelming', or 'distressing' circumstances at the birth.[58] Mary K.'s death sentence was commuted because of concern for her physical and mental health. Reference was also made to Elizabeth H.'s 'weak intellect', which was 'further clouded by the distressing circumstances attending the birth of the child'.[59] In the case of sisters Elizabeth and Rose E., reference was made both to the circumstances of the case and the sex and age of the convicts. From the available evidence it is not clear to what extent medical professionals influenced the outcome in these cases.

During the years 1922 to 1950, barristers defending women charged with the murder of their new-born infants consistently referred to child-birth as a traumatic experience. They argued that because of the physical and mental toll childbirth took on women who had concealed their pregnancies, they could not be held fully responsible for their actions. The statements doctors made regarding the mental state of unmarried mothers who committed infanticide may have helped shape the kinds of arguments barristers used to defend their clients in court. For instance, in his depositional evidence Dr Cecil Molony said that Bridget B., who was charged with the murder of her unnamed male infant in June 1935, 'appeared to be very distressed at what had happened ... She appeared to have suffered greatly both physically and mentally'.[60] Similarly, in a 1939 county Tipperary case, Dr John Murnane said that '[he] formed the opinion that owing to the birth and the attendant anxiety Mary B. was mentally upset'.[61] Doctors who treated unmarried mothers charged with infanticide, referred to the mental disturbance, mental strain, distress, shock, nervousness or suicidal tendencies their patients experienced in their depositions, in 19 cases out of a total of 195 cases. It is quite prob-able that many more doctors referred to these symptoms when called on to give evidence in court and the number of doctors who observed signs of mental anguish in women charged with infanticide may actually be higher.

In three of the 12 cases where a woman was sentenced to death for the murder of an infant, it was a close female relative of the birth mother who was convicted of murder. While barristers defending unmarried mothers charged with infanticide could argue that they could not be held responsible for their actions because they were distressed and mentally unbalanced, if a woman other than the birth mother was charged with infanticide, the notion of shame was usually invoked to defend them. In the Mary S. case the defence lawyer referred to 'the shame which was brought upon householders by the birth of an illegit-imate child'.[62] In defence of his client he argued that 'there was no calm intention of taking away the life of the child' and that the overwhelming sense of shame was 'a natural thing to try to avoid'.[63] The reluctance to execute unmarried mothers who committed infanticide also extended to women other than the birth mother and to men who committed infanticide.

Some doctors drew attention to the fact that young single women suspected of murdering their infants had very little knowledge of pregnancy or childbirth. Like many single women charged with the murder of their illegitimate infants in this period, Mary Anne M. had neglected to cut the umbilical cord. Dr Annie Keogh stated that it was likely that Mary Anne had been in 'great physical and mental agony'. 'A young girl', like Mary Anne, she stated, 'would not have known what to do'.[64] The medical officer who examined the defendant in a 1948 county Kilkenny case remarked that 'from examination it appears that this girl was completely ignorant of the dangers connected with child birth and also with some of the essential physical facts'.[65]

Temporary insanity

The issue of unmarried mothers and temporary insanity is central to any discussion of infanticide and medicine. McAvoy has argued that with the passing of the Infanticide Act in 1949, 'infanticide came to be seen as the act of a woman temporarily mentally unbalanced by the trauma of childbirth'.[66] The idea that following childbirth a woman could become temporarily insane was not new. Cooper Graves has argued that in England 'the idea of "temporary insanity" became increasingly accepted and more commonly used as a medico-legal concept at the turn of the twentieth century'.[67] According to Marland, 'puerperal insanity began to be used with increasing authority as a defence plea in cases of infanticide or concealment' from the early nineteenth century onwards,[68]

> following the trial of giving birth, involving intense physical effort, pain and disruption of the delicate reproductive organs, and in many cases increased strain on family resources, mothers were seen as liable to become deranged, neglectful or violent. They came to represent a risk to themselves and other family members, particularly the new-born child.[69]

Women who suffered from puerperal insanity were regarded as either having been 'unaware of what they were doing or to have temporarily lost control of their actions'.[70] In Ireland, however, while doctors who were called on to give evidence in infanticide trials between 1922 and 1949 frequently noted that single women who had given birth secretly and without assistance may have been dazed, distressed or suffered from shock, few referred in their evidence to the fact that women may have suffered from puerperal insanity or puerperal mania.

In Ireland, doctors had referred to temporary insanity in court long before the law made a provision for it and, in their statements, a number of women who were charged with the murder of a new-born infant described themselves as mad or half-mad immediately after giving birth. In a 1944 county Kilkenny case, Bridget D. told Gardaí that she could not tell them exactly what she did. She said she was 'half mad at the time'.[71] Some women became suicidal. However, statements made by women when they were arrested and charged clearly need to be treated with caution. When Bridget D. said she was 'half mad' she may have merely meant that she had been extremely distressed and emotional. Dr Patrick Hamilton, meanwhile, had 'very considerable experience of maternity cases'.[72] He was of the opinion that 'it is quite common in a first case for the mother to be for a short time in a state of temporary insanity. It is more marked in the case of a first birth'.[73] Dr Charles Conway examined Susan D. after the birth of her new-born infant son in May 1935. He too was of the opinion that 'a woman unattended at the birth of her first child would be mentally disturbed and would not be normal'.[74] In fact, he went as far as to suggest it was common for a first time 'mother to be willing to inflict injury on her child immediately after birth'.[75]

From the depositional evidence of doctors, nurses, family members and the statements of unmarried mothers, it is clear that in many cases women charged with infanticide displayed signs of distress and irrational behaviour. In their statements to the Gardaí, a number of women who were charged with the murder of a new-born infant maintained they were suffering from the shock and trauma of giving birth without assistance and, as a result, were unable to account for their actions. Some women claimed that they were unable to recall what had happened immediately after the birth. Joan C. was charged with the murder of her unnamed male infant in March 1944. When Dr William O'Connor examined the baby's body, he found a knotted cloth tied around the infant's neck. Part of the infant's gum was torn. The doctor concluded that the baby had been strangled and that death was caused by asphyxia. In her statement Joan C. said that she did not recall tying a cloth around the infant's neck and claimed that if she had done so it had not been intentional. Joan insisted that she was suffering from a loss of memory and could not remember what had happened or what she had done. 'I don't remember putting any cloths around the child except my coat ... I don't remember being brought to Tralee hospital'.[76] It is clear from the depositional evidence of Joan's sister and father that they were alarmed by her behaviour. Mary C. gave a very vivid description of her sister Joan shortly after she had given birth. 'She appeared to be convulsed. She was tearing the bedclothes. She was working her hands. As a result of that convulsed condition I brought in her father and he had to hold her in bed'.[77] Joan's father said that 'she was cracking her head against the side of the bed' and 'throwing her hands and gnashing her teeth during all that period'.[78] Joan C.'s behaviour is one of the more extreme examples of mental disturbance that manifested itself in single mothers accused of infanticide.

In a report on a 1931 county Clare case in the *Irish Times*, it was noted that when Margaret F. gave evidence she said that 'before she gave birth to the child in the field she was in such agony that she did not know what happened. Bridget C. gave her a bootlace and told her to tie it around the infant's neck, which she did, because she did not know then what she was doing. She denied having deliberately killed the child'.[79] The jury took this into account and, although they found her guilty, they strongly recommended Margaret F. to mercy 'owing to the circumstances under which the birth took place'.[80]

The jury in the Deborah S. case found it difficult to deliver a verdict. They informed the judge 'that their difficulty was as to the state of mind of the woman at the time of the act'.[81] Deborah S. gave birth on a mountainside in county Kerry in February 1929 without any assistance. The infant's body was found in a stream where it had been drowned. The doctor who carried out the post-mortem in the Deborah S. case stated that 'in the absence of any assistance it is possible but highly improbable for a mother to help herself by putting a hand on the child's throat. In moments of great pain it is possible the mother might do the improbable thing'.[82] Deborah S.'s defence argued that she felt so weak and frightened after giving birth alone that she put the child's body in a bush but could not explain how it had drowned. The jury in the Deborah S. case found the prisoner guilty but recommended her to mercy. 'In all the circumstances of the case, I strongly recommend the accused to mercy – the act was committed immediately after birth when the accused must have still been suffering from the pangs and subsequent prostration of child-birth.'[83]

In his deposition Dr Sean Travers, who treated Catherine A., explained that while he had never seen an infant injured during birth in the way that Catherine's infant may have been injured and killed, he had 'heard and read of such cases'.[84] He believed it was possible that the injuries to Catherine A.'s infant were 'caused by the mother without knowing what she was doing'.[85] The professional opinion of men like Dr Travers was extremely important in infanticide cases. Doctors formulated arguments that linked childbirth to mental disturbance and, in doing so, they provided a loophole that made it easier for the government and the Governor General to commute the death sentence passed on women who were convicted of the murder of their new-born infants.

The statements of unmarried Irish mothers who were charged with infanticide between 1922 and 1950 suggest that many women may have convinced themselves that the infant they were due to deliver would be dead. A number of women described the acute sense of distress they experienced when they realised that the infant was alive and because they had not prepared for such an outcome, they generally strangled or suffocated the infant almost immediately. Bridget C. initially told Gardaí that 'the child was dead when it was born'.[86] However, she later admitted that the baby had been alive. She explained that she was distressed when she realised that the infant was alive. 'I will tell you the truth. I was very upset the child was born alive. I held it in my hands for about half an hour and I think it died with the cold. I was so upset I was not sure if it was dead when I put it in the haycock.'[87]

The language that unmarried mothers charged with infanticide used to refer to their infants suggests that many women viewed their infants extremely negatively. They saw them as a terrible mistake or a shameful secret that had to be concealed and destroyed. Bridget D. referred to her baby as 'the thing'.[88] Margaret F. strangled her new-born infant with a bootlace in June 1930. When her cousin asked her if it was 'a sin to [kill] it, without being baptised' and remarked that the baby 'would be in darkness all its lifetime', Margaret simply said, 'let it, the bastard'.[89]

Case 163 stands apart from the vast majority of infanticide cases in this sample, as Eithna M.'s illegitimate son was seven months old when he was drowned by his mother in September 1940. In her statement, the 19 year old Eithna told Gardaí that on Tuesday 17 September 1940, she 'made up [her] mind to drown the child in Gurtavehy lake as [she] [had] been annoyed from it since it was born'.[90] She spent the entire day at the edge of the lake after she had killed her son. According to Eithna's second cousin, with whom she and her baby lived, two days before the incident, when asked to attend to her baby, Eithna had remarked 'Am I going to be all my life watching the baby?'[91] Eithna's behaviour suggests that she may have been suffering from a form of depression but, as there are no medical reports in her file, this is by no means certain. She may simply have resented the responsibility of motherhood and deliberately decided to take her son's life.

Although a number of women referred to the fact that they had experienced mental distress and one woman claimed she was half mad following the birth of her infant, very few single women convicted of the murder, manslaughter or concealment of birth of their illegitimate infants, were actually found insane. Between 1922 and 1950 a total of three single women for whom information is available (1.9 per cent), were deemed insane and considered unfit to plead. Annie C. was found insane and considered incapable of pleading in June 1930. She was sentenced to be detained in strict custody and taken care of 'until the pleasure of His Excellency the Governor General be made known concerning her'.[92] At her trial for the murder of her twin female infants in February 1930, Kate M. was found guilty but insane at the time. Margaret F. was charged with the murder of her illegitimate male infant in June 1943. The doctor who had been called to Margaret's house after she had given birth said that she was 'slow witted', and the medical officer who examined Margaret in Sligo prison was of the opinion that she was mentally defective.[93] Margaret was found insane and ordered to be detained indefinitely in strict custody. Elizabeth D. was sentenced to death, but the sentence was later commuted to penal servitude for life. She may later have been deemed insane and in need of psychiatric care as she was transferred to Dundrum Asylum in July 1926. As detailed medical reports do not exist for these individuals, it is not known how their insanity manifested itself.

Suicide

A total of 11 women told family members, the Gardaí, their doctors, or the medical officers who examined them in prison, that they had considered taking their own lives. In addition, one woman was charged both with concealment of birth and with attempted suicide. Johanna D. attempted to commit suicide in August 1926 by cutting her throat with a razor. The entry in the Trials Record Book from June 1925 to June 1927 indicates that Johanna attempted suicide almost a week after she concealed the birth of her infant. Johanna told Guard Kennedy that 'she wanted to cut her throat, that she had her people scandalised'.[94]

On 15 January 1935 a parcel containing the body of a child was found in a disused quarry in county Limerick. Margaret D. was arrested and charged with infanticide. She told the Gardaí that '[she] was distracted and mad and ... could not stick it at the time'.[95] In his deposition her father stated that '[he] heard [his] daughter say that if she had a razor she would cut her throat'.[96] Mary C. was tried for the murder of her unnamed male infant in June 1947. She repeatedly threatened to commit suicide in Sligo prison. The medical officers who treated her in Sligo prison took a number of precautions to prevent her from taking her own life. On 4 May the medical officer in Sligo prison came to the conclusion that Mary had suffered from a 'brain storm' after giving birth.[97] The medical officer ordered Mary to be placed under complete observation 'in view of the fact that she was capable of suffering a brain storm' and added that 'all utensils be removed from her cell' and 'a night guard should be posted during her detention'.[98] On 10 May, Mary was recommitted on remand from Letterkenny district court and she told the prison doctor once again 'that she did not wish to live'.[99] The medical officer recorded that she was to be kept 'under strict observation day and night, with a view to preventing her from committing suicide or injuring herself'.[100] However, on 17 May, a week later, Mary assured the medical officers that 'she had at no time any intention of doing away with her life'.[101] The medical officers concluded that 'she is a person of sound mind'.[102]

Johanna D. clearly attempted to end her own life and Mary C.'s threats were taken very seriously by the prison authorities. Mary C. was interviewed on a number of occasions and a number of measures were taken by the prison authorities to ensure that she could not commit suicide. However, it is not clear in all cases whether women charged with the murder of their new-born infant actually attempted to commit suicide. It is also difficult to assess how serious their threats were.

'Half-mad at the time'?

Margaret D. said she was distracted and mad following the birth of her baby and Bridget D. said she was 'half mad at the time'.[103] Several women in this sample claimed that they could not remember what happened immediately after self-delivery. A number of doctors expressed their concerns regarding the mental health of their patients in their depositions. However, it is only possible to speculate about how many women may actually have experienced temporary insanity. While some women in this sample may well have experienced temporary memory loss following the birth of their illegitimate infants, others may have claimed that they were unable to recall what had occurred in the hope that they might be dealt with more leniently. The deaths of some infants may have been unintentional but many more may have been deliberately killed by their mothers.

Many of the infants in these cases were killed in quite brutal ways. What drove young Irish women to such extreme lengths? While some women undoubtedly experienced some form of mental disturbance, others may have been motivated to kill their infants by feelings of shame or fear or by their inability to provide for their infant. Single mothers understood that the birth of an illegitimate child would have an adverse effect on their own lives. Unmarried domestic servants who became pregnant would have lost their jobs and, without a reference, would have found it very difficult to secure a position again. These women belonged to a society that condemned unmarried motherhood without qualification and offered little support for single mothers and their children. Most unmarried mothers had little or no financial means and would have been unable to support a child. Without a social welfare system for impoverished single mothers to fall back on, and without the support of the infant's father, it is not difficult to see what may have motivated many ordinary young women to conceal the births of their illegitimate children and to kill them.

Conclusion

As McAvoy has argued in her study on infanticide in the Irish Free State, the state authorities employed 'a complex range of legal strategies' in order to ensure that infanticide was treated differently from other categories of murder.[104] With the passing of the Infanticide Act in 1949 a woman charged with infanticide was no longer considered fully responsible for her actions. It was assumed that she had been suffering from a disturbed or unbalanced state of mind at the time and infanticide became a non-capital offence. However, even before the Infanticide Act was passed, barristers defending unmarried mothers charged with infanticide argued that giving birth alone without any assistance was such a traumatic ordeal that they may have been temporarily mentally imbalanced and, as a result, may have unintentionally harmed their infants. The introduction of the Infanticide Act in 1949 did not mark a significant change in the way the courts dealt with women convicted of the murder of their infants. Although many doctors who had treated unmarried women charged with infanticide referred in their depositions to the kind of abnormal mental and nervous condition that some women experienced as a result of childbirth, the medical profession seem to have had little influence in the drafting of the 1949 Infanticide Act. McAvoy has noted how the legislation relating to infanticide was 'formulated by a small group of civil servant and politicians without medical advice'.[105]

Many doctors were sympathetic towards the unmarried women they treated who had been charged with infanticide during the years 1922 to 1950. In many cases, doctors who treated them did not wish to see such young women convicted of murder. Doctors who had treated unmarried mothers charged with infanticide and carried out the post-mortem examinations on their infants almost always confirmed that the infant had been killed but their statements describing the cause of death were often followed by observations regarding the mental state of the unmarried mother at the time the act was committed. Some doctors may have been unaccustomed to treating women who were traumatised and suicidal. This is reflected in the insensitive remarks made by a small number of doctors. For the most part, nurses do not feature as prominently as doctors in the records of infanticide trials, although they were also actively involved both in the care of unmarried mothers and in the courtroom.

The alleged criminal behaviour of nurse Mary Anne Cadden stands in sharp relief to the kind of care provided by most doctors and nurses during the period under review. However, the treatment of unmarried mothers in county homes highlights the fact that there were clear limits to the kind of care offered by staff in Ireland's network of county homes. Unmarried mothers were warned to look after their infants but were given little in the way of practical help or advice. There is evidence to suggest that members of some hospital committees and county health boards were hostile towards single mothers who sought admittance to maternity wards. In June 1927 at the monthly meeting of the Galway hospital and dispensaries' committee, the members of the committee were informed that five unmarried mothers had recently been admitted to the hospital. The committee was clearly dissatisfied and a resolution calling on members of the Catholic hierarchy to appeal to the people of Ireland to 'return to the old Gaelic customs under which such scandals were practically unknown' was passed unanimously.[106] In May 1929 at a meeting of the County Wexford health board Miss O'Ryan referred to unmarried mothers as 'people who don't deserve respect'.[107] She argued that 'respectable labourer's wives',[108] who required treatment in the county home during their confinement, should not have to 'be put into the same ward as the unmarried mother'.[109] A number of others in attendance at the meeting clearly agreed. Mr Gaul said that he felt that 'there should be some provision made in the County Hospital for decent married women during their confinement, who could not think of going into the County Home amongst unmarried mothers'.[110] As the author of an article on illegitimacy in *The Bell* in 1941 suggested, if the number of infant deaths in Ireland were to be reduced, social services would have to be developed and improved but perhaps, more fundamentally, people's attitudes towards unmarried mothers and their children would have to change.[111] Judging by the remarks made by members of hospital committees and county health boards in the 1920s, it is clear that hostility towards unmarried mothers was deeply ingrained in Ireland. Such widespread negative attitudes meant that social services for single mothers and their children did not improve until much later in the twentieth century.

Notes

1   I would like to thank the Irish Research Council for the Humanities and Social Sciences for funding the research on which this chapter is based.
2   The full names in cases quoted in this chapter are available in the original material.
3   Central Criminal Court (hereafter CCC), ID 24 131, county Leitrim, 1944, National Archives of Ireland (hereafter NAI).
4   Ibid.
5   Cath Quinn, 'Images and impulses: Representations of puerperal insanity and infanticide in late Victorian England', in Mark Jackson (ed.), *Infanticide: Historical Perspectives on Child Murder and Concealment, 1550–2000* (Aldershot, 2002), p. 202.
6   Margaret L. Arnot and Cornelie Usborne, 'Why gender and crime? Aspects of an institutional debate', in idem. (eds), *Gender and Crime in Modern Europe* (London, 1999), p. 20.
7   CCC, ID 11 95, county Donegal, 1940, NAI.
8   *The Irish Catholic*, 16 February 1929.
9   Ibid.
10   Ibid.
11   CCC, ID 24 131, county Leitrim, 1944, NAI.
12   Ibid.
13   CCC, ID 33 65, county Limerick, 1932, NAI.
14   Ibid.
15   Bridget C. was a married woman. Although the focus of this study is on single women who were charged with infanticide, I have referred to the Bridget C. case here as an example of the ways in which the Gardaí may have intimidated women suspected of infanticide.
16   Court of Criminal Appeal (hereafter CCA), no. 31 of 1934, NAI.
17   Ibid.
18   Ibid.
19   Lindsey Earner-Byrne, 'The boat to England: An analysis of the official reactions to the emigration of single expectant Irishwomen to Britain, 1922–1972', *Irish Economic and Social History*, 30 (2003), p. 69.
20   Finola Kennedy, *Cottage to Crèche: Family Change in Ireland* (Dublin, 2001), p. 145.
21   Ibid.
22   Mary Raftery and Eoin O'Sullivan, *Suffer the Little Children: The Inside Story of Ireland's Industrial Schools* (Dublin, 1999), p. 73.
23   *Clare Champion*, 25 November 1933.
24   Ibid.

25 Ibid.
26 *Irish Times*, 22 January 1926.
27 *Irish Times*, 26 November 1930.
28 Ibid.
29 CCA, No. 24 of 1930, NAI.
30 Ibid.
31 Department of Justice, H247/41D, 90/4/8, NAI.
32 Ibid.
33 *Irish Times*, 27 March 1935.
34 CCC, ID 11 93, county Tipperary, 1939, NAI.
35 CCC, ID 24 131, county Kerry, 1944, NAI.
36 Mary Anne Cadden ran a private nursing home in Rathmines in Dublin and Florence Lang ran a nursing home in Cavan.
37 CCC, ID 22 83, county Galway, 1943, NAI.
38 CCC, ID 11 94, county Meath, 1939, NAI.
39 CCC, ID 11 95, county Dublin, 1939, NAI.
40 Ibid.
41 Ibid.
42 Ibid.
43 *Irish Times*, 13 May 1939.
44 Trials Record Book, ID 11 92, NAI.
45 CCC, ID 56 30, county Meath, 1936, NAI.
46 Department of the Taoiseach, S 5886, NAI.
47 CCC, ID 29 7, county Kilkenny, 1948, NAI.
48 Ibid.
49 CCC, ID 11 96, Dublin, 1941, NAI.
50 CCC, ID 60 58, county Donegal, 1937, NAI.
51 CCC, ID 29 11, county Tipperary, 1949, NAI.
52 Ibid.
53 CCC, ID 27 5, county Sligo, 1945, NAI.
54 CCC, ID 11 29, county Sligo, 1949, NAI.
55 CCC, ID 24 131, county Kilkenny, 1944, NAI.
56 Department of the Taoiseach, S 7788A, NAI.
57 Sandra McAvoy (Lamour), 'Aspects of the state and female sexuality in the Irish Free State, 1922–1949' (PhD thesis: University College Cork, 1998), p. 310.
58 Department of the Taoiseach, S 7788A, NAI.
59 Ibid.
60 CCC, ID 56 22, county Limerick, 1935, NAI.
61 CCC, ID 11 93, county Tipperary, 1939, NAI.
62 CCA, no. 36 of 1938, NAI.
63 Ibid.
64 CCC, IC 90 27, county Wexford, 1927, NAI.
65 CCC, ID 29 7, county Kilkenny, 1948, NAI.
66 McAvoy, 'Aspects of the state and female sexuality', p. 276.
67 Donna Cooper Graves, '"In a frenzy while raving mad": Physicians and parliamentarians define infanticide in Victorian England', in Brigitte H. Bechtold and Donna Cooper Graves (eds), *Killing Infants: Studies in the Worldwide Practice of Infanticide* (Lampeter, 2006), p. 130.

68 Hilary Marland, 'Getting away with murder? Puerperal insanity, infanticide and the defence plea', in Jackson (ed.), *Infanticide: Historical Perspectives*, p. 174.
69 Ibid., p. 173.
70 Ibid., p. 172.
71 CCC, ID 24 131, county Kilkenny, 1944, NAI.
72 CCC, IC 90 28, county Wicklow, 1926, NAI.
73 Ibid.
74 CCC, ID 56 18, county Dublin, 1935, NAI.
75 Ibid.
76 CCC, ID 24 131, county Kerry, 1944, NAI.
77 Ibid.
78 Ibid.
79 *Irish Times*, 4 March 1931.
80 Ibid.
81 Department of the Taoiseach, S 5886, NAI.
82 CCC, IC 94 61, county Kerry, 1929, NAI.
83 Department of the Taoiseach, S 5886, NAI.
84 CCC, IC 94 75, county Wexford, 1929, NAI.
85 Ibid.
86 CCC, ID 11 93, county Longford, 1939, NAI.
87 Ibid.
88 CCC, ID 24 131, county Kilkenny, 1944, NAI.
89 CCC, IC 94 54, county Clare, 1930, NAI.
90 CCC, ID 11 96, county Cork, 1941, NAI.
91 Ibid.
92 Trials Record Book, CCC, ID 33 68, Change of Venue Cases, November 1927 Sittings – June 1933 Sittings, NAI.
93 CCC, ID 22 84, county Mayo, 1943, NAI.
94 *Limerick Leader*, 22 September 1926.
95 CCC, ID 56 22, county Limerick, 1935, NAI.
96 Ibid.
97 CCC, ID 29 1, county Donegal, 1947, NAI.
98 Ibid.
99 Ibid.
100 Ibid.
101 Ibid.
102 Ibid.
103 CCC, ID 24 131, county Kilkenny, 1944, NAI.
104 McAvoy (Larmour), 'Aspects of the state and female sexuality', p. 277.
105 Ibid., p. 319.
106 *Irish Times*, 20 June 1927.
107 *People* (Wexford), 25 May 1929.
108 Ibid.
109 *People* (Wexford), 22 June 1929.
110 *People* (Wexford), 25 May 1929.
111 M.P.R.H., 'Illegitimate: being a discourse on the problems of unmarried mothers and their offspring', *The Bell*, 2:3 (1941), p. 87.

9
Venereal Disease in Interwar Northern Ireland

Leanne McCormick

Introduction

The first comprehensive attempt to discover the prevalence of venereal diseases (VD) in the United Kingdom took the form of a Royal Commission on Venereal Disease (RCVD), which was established in 1913 and made its report in 1916. The final report of the RCVD recommended that centres should be set up for the treatment of VD, with 75 per cent of the cost to be met by central government and the remainder from local rates.[1] The establishment of clinics relied on cooperation between local authorities and voluntary hospitals, but this was to prove a problematic relationship in Northern Ireland,[2] as it had done in other parts of the United Kingdom.[3] The situation in Northern Ireland, however, was exacerbated due to the complex political situation, particularly the establishment of the new state of Northern Ireland. The 1920 Government of Ireland Act partitioned Ireland and set up two governments and two parliaments, one for the six counties that were to form Northern Ireland and another for the 26 southern counties that became the Irish Free State. The Anglo-Irish treaty, which was signed in December 1921, brought to an end the three-year conflict between the British forces and those fighting for independence. The treaty gave Ireland the status of a dominion within the British Commonwealth and established the Irish Free State. Northern Ireland was permitted to opt out of the agreement and retain its status granted in 1920, to remain part of the United Kingdom.

Along with the associated violence and upheaval of the period, local government was being reorganised and that added new complications to attempts to enforce legislation. A similar situation existed in the Irish Free State with VD treatment schemes also being adversely affected by the turbulent political situation there.[4] The enforcement of VD legislation in Northern Ireland and the debates and difficulties surrounding this legislation have been largely neglected,[5] as recent research has largely focused on the 26 southern counties of what is now the Republic of Ireland.[6] This article hopes to partially redress the balance by considering the issues involved with the establishment of treatment schemes and centres in Northern Ireland, the propaganda and publicity about VD in the interwar years, and the changing attitudes towards the diseases during this period.

Implementing a VD scheme

The interwar years were marked by a general lack of enthusiasm and reluctance to engage with the problem of VD in Northern Ireland. Following the recommendations of the Royal Commission on VD in 1916, councils were required to submit proposals for VD treatment schemes to the Local Government Board (LGB). These met with opposition from the six councils of counties Antrim, Armagh, Down, Fermanagh, Londonderry[7] and Tyrone, which were to become Northern Ireland. By early 1919 treatment centres had only been established in Antrim, Armagh and the County Borough of Belfast.[8] Many patients from outside these areas travelled to Belfast for treatment, which led to an inevitable debate as to which council was responsible for their treatment expenses. Fermanagh County Council, for example, did adopt a VD treatment scheme in February 1919, but the LGB rejected it as the council had not set up treatment facilities in the county and patients were instead going to Belfast to be treated.[9] Following a claim by the Belfast Corporation in 1920 to the LGB, concerning payment for patients from Fermanagh treated for VD in the Royal Victoria Hospital in Belfast (RVH), it was decided that patients from there would be counted under the Belfast scheme.[10] It was not until 1923, when the new Ministry of Home Affairs took over responsibility for implementing the VD treatment schemes, that increased pressure was put on Fermanagh County Council to adopt a scheme, 'however tentative'.[11]

Fermanagh County Council was not alone in its reluctance to establish VD treatment centres. The debates that occurred in county Tyrone about the implementation of a treatment scheme illustrate very clearly many of the contemporary fears and objections surrounding the treatment of the diseases. One of the most commonly expressed views was that the proposal to set up treatment centres was not a matter for county councils, but was the responsibility of the LGB and central government or medical personnel. As Councillor Murnaghan, who was one of the most vocifer-ous opponents of the scheme, argued, accepting the proposal meant that a council, mainly composed of business men, should take on the man-agement of a disease they were told was highly infectious. Moreover, he was indignant at being

> asked to make provision for the treatment of those degraded people who had fallen lower than the brute beasts, and to take in hand the curing of these people partly at the expense of the honest, upright rate paying public of the county. They had no right to treat these people; they had no means of treating them ... this was not a matter for the County Council, but for the Local Government Board to deal with in a similar manner as the Department of Agriculture deal with serious diseases in animals, by taking hold of them and insisting on the disease being stamped out.[12]

At a meeting of the Tyrone Council Hospital Committee in December 1918, Mr Murnaghan had raised similar arguments and stated that the hospital was 'for the ordinary, clean, decent people of the county' and that VD should be treated 'as you do all dangerous and virulent diseases – by isolation'.[13] The idea that ordinary, clean, decent people would be contaminated and put at risk by treating VD in local hospitals was widespread. Monsignor Doherty, a Catholic priest and Chairman of the Council Committee, expressed a similar opinion, that the County Hospital was built for clean patients and that he felt it was 'intolerable to think or talk about it all. It really is a shame that men should talk about it'. He also reiterated a commonly held contemporary view that associ-ated VD with the army rather than the general population, and argued that 'the Government have plenty of military hospitals for these abnor-mal cases, and they should not introduce them into this institution [Tyrone County Hospital] as the people know nothing about it'.[14]

However, the belief that the general public was completely ignorant about VD, and that it was not prevalent in Northern Ireland, was rejected by Dr Thompson, the medical representative on the Tyrone County Hospital Committee. In answer to Mr Murnaghan's arguments he stated that not only was it a medical matter but that 'if the people of the country were all angels, as Mr Murnaghan thinks they are, it would be alright!'.[15] He argued that VD was much more prevalent than the Committee members realised and concluded the debate by stating that 'if any person suffering from a moral or immoral disease comes to me for treatment I would treat them. That is what I am there for.'[16]

There was clearly little support within County Tyrone for the proposed voluntarist approach to the treatment of the diseases. It has been argued that the British voluntarist approach reflected a demand to retain local authority over the treatment of VD.[17] However, it is clear from the discussions of Tyrone County Council that, as far as VD was concerned, centralisation, direct state intervention and regulation were preferred. Those who wished to retain local control regarded this as a means to avoid adopting the schemes rather than implementing them, as Mr Donnelly, a member of Tyrone County Council, explained: 'we would be much better satisfied if the LGB would take the matter out of our hands and deal with it themselves'.[18] Nonetheless, in November 1919 a scheme was passed, albeit reluctantly, by Tyrone County Council. There was still evidently no great swell of support for the scheme but rather, there existed a general acceptance that legislation had to be passed.

Londonderry County Council also resisted implementing a treatment scheme. Its VD committee minutes contain only two entries, one from 1918 postponing discussion until the medical profession considered the idea, and the second from 1919 further adjourning the committee.[19] It was not until the 1940s that treatment centres were actually established in the county and borough.[20] The disruption experienced in Ireland, north and south, in the early 1920s undoubtedly contributed to the ability of councils like Londonderry's to avoid implementing an effective treatment scheme. The partition of Ireland and the creation of Northern Ireland necessitated the formation of a new system of government and, in terms of public health, a redistribution of the responsibilities of the LGB. In Northern Ireland, the Ministry for Home Affairs took on responsibility for health and, in the Irish Free State, a Department of Local Government and Public Health was established. In the midst of reorganisation and the violence associated with partition, it is unsurprising that a controversial issue such as VD was not the focus of attention.

Nonetheless, by 1922, treatment centres for VD were operational in Belfast, as well as in nine towns across counties Antrim, Armagh, Down and Tyrone. Separate clinics were run for men and women and also for the treatment of syphilis and gonorrhoea.[21] Within Belfast there were clinics at the RVH, where the patients could enter from a separate door in an effort to preserve confidentiality, the Mater Hospital, and also the Belfast Union Hospital. The Union Hospital, which was attached to the workhouse and run by a Board of Guardians, had previously treated patients in male and female 'lock wards', which were isolated wards for the treatment of VD. As these wards were now under the VD treatment scheme, it was decided to make them more open and to rename them as 'Ward 6' for men and 'Ward 26' for women to try to remove the stigma attached to them and encourage more people to come for confidential treatment.[22]

Financial complications

As referred to above, one of the first problems involved in operating the VD treatment scheme was financial. County councils without adequate treatment centres often sent patients to be treated in Belfast. The problem for the Belfast Corporation was collecting payment for the treatment of patients coming from outside the city. This was a problem also experienced in Scotland, where city authorities feared the influx of patients from rural areas.[23] Although the Belfast Corporation had agreed terms on which patients from counties Antrim, Fermanagh, Londonderry and Tyrone could be treated in Belfast in 1919,[24] the system appeared to have broken down by 1922.[25] This breakdown occurred as a result of the transference of powers from the LGB to the Ministry of Home Affairs, because the arrangements made to 'apportion expenses' had been overlooked in the transition and the Corporation now felt that it was 'highly desirable to arrive at some definite standing with the Ministry of Home Affairs'.[26] It is unclear whether this was definitely resolved as the Corporation appealed again for action by the Ministry of Home Affairs in 1926.[27]

The complications of such a system are illustrated in the correspondence that took place between the Executive Sanitary Officer for Londonderry, Mr Fletcher, and the Belfast Corporation's Accountant, Mr Gearle, in 1923.[28] Mr Fletcher explained that an auditor had called for proof that the £25, 10s., 11d., that had been paid to the Belfast Corporation was definitely for patients from Londonderry, and asked if the names and addresses of these patients could be forwarded to the Londonderry Medical Superintendent of Health (MSOH) to be verified by the auditor.[29] In response Mr Gearle pointed out that under the LGB's regulations of 1917, it was prohibited to disclose the names and addresses of patients. He closed the letter rather sharply by suggesting that 'you draw the Auditor's attention to these Regulations, and I may mention that I am communicating privately with the Ministry of Home Affairs'.[30]

The conflict between local authorities and voluntary hospitals in Scotland in the implementation of the VD scheme, over issues such as accommodation, hospital autonomy and status, did not exist to the same extent in Northern Ireland.[31] In many cases, the hospitals outside

Belfast were not implementing the schemes at all and the majority of patients were treated in Belfast. Here the areas of most contention appeared to be between the Corporation and the hospitals over finance and the money available for paying staff and funding treatment. The Corporation Minutes record regular requests for increased doctors' salaries and funding for the clinics. The Corporation was reluctant to increase funds for these requests and in 1932 the Public Health Committee even proposed that charges should be made for the drugs used to treat VD.[32] The Ministry of Home Affairs was, however, quick to point out that under the 1917 regulations, the drugs had to remain free.[33]

Through the interwar years, the Northern Ireland government was financially strapped and the amount of money that was spent on health and social services hardly changed over the whole period.[34] Local authorities were also reluctant to spend and it has been argued that the conservative nature of local government caused inactivity and a lack of desire to improve conditions.[35] The Belfast Corporation was not only parsimonious but also suffered several corruption scandals during and just after the interwar years: a 1926 report revealed corruption in the Housing Committee, and in 1941 the Corporation Tuberculosis Committee was dismissed over mismanagement of the Whiteabbey Sanatorium.[36] With wide financial constraints and a reluctance to cause controversy or provoke public reaction, it is unsurprising that there was not only an unwillingness to financially support VD treatment centres but also to advertise their existence.

Publicity programme

With the establishment of a VD scheme came the question of publicity and who was to organise it. In keeping with the recommendations of the RCVD, responsibility for communicating to the public the dangers of VD rested with the National Council for Combating Venereal Disease (NCCVD). The Ulster Branch of the NCCVD was very keen to be involved in publicity and propaganda from the outset. A branch had been established in 1917 and representatives, headed by Prof. J.A. Lindsay,[37] had been in talks with the Corporation Public Health Committee about the question of publicity. They had recommended in 1918 that when making their claim for 75 per cent of the cost of the VD scheme to the LGB, they include an application from the NCCVD for £300 for publicity.[38] This was to pay for lecturers and also to obtain the services of an organising secretary.[39]

It was not until 1920 that an organising secretary was actually appointed. It was felt that the job would be suitable for an ex-serviceman and a Mr Garner was appointed in January 1920.[40] However, his appointment was short-lived and he left to take up another position in September 1920.[41] Some of the reasons for this short-lived post are revealed in the correspondence between Dr Bailie, the Belfast MSOH, and Mrs Neville-Rolfe, Secretary of the NCCVD in London.[42] There was a flurry of letters between the two from December 1919 to June 1920 concerning the appointment of Mr Garner and the plan to send him to London for training. However, this training did not happen and correspondence between Belfast and London appeared to cease, until April 1921 when Mrs Neville-Rolfe wrote to that say she had been informed that the activities of the Ulster Branch of the NCCVD had come to a halt, and to ask what had happened concerning Garner's training.[43] She goes on to say, 'I presume that the political conditions are such that very little educational work can be undertaken, but in all probability those very conditions of unrest favour a spread of VD'.[44] In reply Dr Bailie explained that as Mr Garner's contract was only confirmed to September 1920, it was decided that it was not worthwhile sending him to London. He praised the work that Mr Garner had done when he was with them.[45] Between May and August 1920 he had visited over 60 workplaces and given 28 talks.[46] Dr Bailie informed Mrs Neville-Rolfe that as the clinics at the RVH and the Mater were very busy, with the clinic at the RVH open six days a week, this large volume of work had led the Public Health Committee to believe that neither a replacement for Mr Garner or any more publicity were necessary.[47]

Evidence that there were perhaps other reasons for the lack of replacement for Mr Garner were revealed in a discussion about VD propaganda in Éire in the 1940s. The Department of Health justified its decision not to publicise VD treatment facilities by referring to the situation that had occurred in Ulster in the early 1920s, where lectures and the distribution of literature about VD had received little support. Following the public objections of Rev Dr McHugh, the Catholic Bishop of Derry, the LGB had apparently discouraged any further plans for lectures or publicity.[48] This highlights the strength of the opposition to anything associated with VD in the county Londonderry authorities and provides an additional explanation as to why Mr Garner might not have been replaced.

The reluctance to carry out propaganda and publicity about the treatment of VD and the dangers of the disease was in marked contrast, for example, to the comprehensive publicity programmes carried out in Scotland.[49] While the same level of publicity did not occur in Northern

Ireland as in Scotland, similar controversies about the displaying of VD propaganda posters in public toilets occurred in both Belfast and Glasgow.[50] The display of notices relating to issues such as VD in public toilets was prohibited under Section 58 of the Belfast Mains Drainage Act of 1887. Dr Bailie, the Belfast MSOH, was refused permission to put posters in public urinals by the Corporation Police Committee in 1919 under the 1887 Act.[51] He made a further attempt to change the act the following year, and wrote to the Corporation Pubic Health Committee to explain how other cities in Britain had put up similar posters.[52] The issue was still not resolved by 1933 as Charles Thomson, who had become MSOH in 1928, again tried to persuade the Public Health Committee of the importance of using such posters. He explained how many cities in Europe used posters in public toilets to draw attention to the dangers of VD and the importance of receiving prompt treatment.[53] It was decided that Dr Thomson should contact the Improvement Committee to try to get the legislation changed. He appealed to them to change what he saw as archaic legislation and argued that public opinion had changed, and it was now felt that 'tactfully worded notices should be placed on such walls with a view to eliminating as far as possible this scourge which has wrought such havoc in the past'.[54]

This appeal clearly did not work as the Public Health Committee minutes for January 1934 recorded a request to the Minister of Home Affairs to approve the placing of notices drawing attention to the dangers of VD in public toilets.[55] Writing to Dr Rankin at the VD department of the RVH in July 1937, Charles Thomson revealed how the legislation still had not changed: 'an old Corporation Act forbids reference in public urinals etc. to VD and centres for treatment. I am prepared to approach my Committee again to try and get an Act passed to amend this'.[56] He did approach the Committee in both 1937 and 1939 but no solution was reached until 1942.[57] It was only following concerns that the rates of VD had risen, following the outbreak of the Second World War, that efforts were renewed to change the legislation. In 1942, Charles Thomson was finally allowed to prepare notices to be displayed in public toilets.[58]

As MSOH for Belfast, Charles Thomson appears to have become increasingly concerned with the issue of VD by the early 1930s, when much greater discussion of the diseases appeared in his Annual Reports.[59] He renewed contact with the NCCVD, now the British Social Hygiene Council (BSHC), in London, to rectify the cessation of correspondence and cooperation that had occurred in 1921. Mrs Neville-Rolfe had written to Charles Thomson to tell him of the film *Damaged Lives* and to encourage him to bring the film to Belfast.[60] *Damaged Lives* was a fictional VD propaganda film, one of a series of such films that were shown across the United Kingdom in the 1920s and 1930s.[61] The film was about a young man who caught syphilis following a one night stand before his marriage. He passed the disease on to his wife, who believed it to be incurable and tried to commit suicide. She was saved by her husband who, in the educational part of the film, had been shown the devastation that VD can cause but had also learned that it was curable.

Initially rather reticent about bringing the film to Belfast, Charles Thomson explained to Mrs Neville-Rolfe how they had just had a very successful health week, which had been attended by over 25,000 people.[62] However, following a letter from Mrs Neville-Rolfe praising the film, he appears to have changed his mind and the *Damaged Lives* was booked for 12 days in the Ulster Hall in September 1934.[63] The Ulster Hall was a large and important venue in Belfast and Charles Thomson was at pains to ensure that Mrs Neville-Rolfe realised this after she had referred to the Ulster Hall as a cinema. This, he explained, was tantamount to suggesting that 'St Paul's Cathedral had been let for a fortnight for rag-time dances', since in Northern Ireland the Ulster Hall was viewed almost as holy ground.[64]

The *Belfast Telegraph* hailed the film as a 'dramatic masterpiece' that combined 'love, pathos, drama and humour, with a subject which a few years ago was only discussed in whispers'.[65] The publicity the film received was largely positive, though it is worth noting that there was no discussion of the film in the *Irish Times*. Moreover, while clergymen of all religions were asked to provide a statement of support for the film, none was received from the Catholic Church, while others, including Rabbi Shachter of the Jewish Synagogue in Belfast and the Rev. Doherty of the Mission to Seamen, did provide statements of support.[66] Rev. Lindsay, Rector of St Bartholomew's in Belfast, also devoted two Sunday sermons to the film and commented that 'with its splendid teaching and its attempt to banish ignorance regarding a dreadful evil, it will do much towards assisting the campaign for suppressing it [VD]'.[67]

Opposition to the film appeared rather belatedly at a Belfast City Council meeting on 1 October 1934. Councillor Kilpatrick objected to the film being shown under the auspices of the Public Health Committee, as 'Belfast still rejoiced in being a religious, God-fearing community and a picture like *Damaged Lives* was an insult to the city'.[68] In response Alderman Ridgley argued that he believed it was one of the finest things the Committee had accomplished and that it was 'a crusade against superstition and a scourge which should have the approval of all sensible and right-thinking people'.[69] Several other councillors raised objections but, as Councillor Lieutenant-Commander Harcourt pointed out, as the council had already given its approval and the film had been shown, the discussion was out of order.[70]

Preventing VD

As mentioned above, Charles Thomson replaced H.W. Bailie as MSOH for Belfast in 1928. A native of Scotland and educated at Glasgow University, he had held a number of MSOH posts in England before coming to Belfast.[71] He had been a Fellow of the Royal Society of Public Health from 1929 and his obituary described how he was well known in public health circles throughout the British Isles.[72] Through the 1930s Charles Thomson became increasingly concerned about VD and the best way to prevent and treat it. The issue of compulsory notification particularly vexed him and he continued to debate it until his retirement in 1945.

In Scotland the issue of compulsion became one of both medical opinion and public debate in the late 1920s and the Edinburgh Corporation went so far as to submit a bill to the House of Commons in 1928, which, if it had been passed, would have instituted compulsory notification and treatment of VD.[73] In Northern Ireland, although there was some discussion from county councils about centralising treatment when the proposals for a treatment scheme were first introduced, there was no public debate and even within the medical community there was very little discussion of the issue. This lack of medical debate extended to the treatment and prevention of VD in general. Only five articles on the treatment of syphilis or gonorrhoea were published in the *Ulster Medical Journal* between 1917 and 1940.[74]

Charles Thomson's discussion of compulsory notification was prompted by the proposal in February 1935, by two members of the Belfast Corporation Public Health Committee, that the way to prevent the spread of VD was to arrest all prostitutes and have them detained for treatment.[75] In a letter to Dr Rankin, head of the VD clinic at the RVH, Dr Thomson referred to a report he had prepared for the Public Health Committee, in which he had 'knocked out the idea of reverting to the Contagious Diseases Acts' believing that this would 'put an end to any demand for locking up the professional women'.[76] Thomson referred to Bradford and Brisbane, as the only two cities where VD was notifiable, and explained how he had been in communication with Dr Buchan, the Medical Officer of Health for Bradford, to ask about the effectiveness of compulsory notification. Dr Buchan had told him that the 'effectiveness of the power of compulsory notification was seriously diminished by the want of compulsory power of treatment'.[77]

Dr Thomson wanted to know Dr Rankin's thoughts on the issue and whether he felt that to introduce compulsory notification would drive VD underground and whether such powers would be useful.[78] In concluding the letter Charles Thomson revealed his own feelings, when he instructed Dr Rankin to speak freely, as 'I am, by no means seeking to give a soft answer to the Committee. Edinburgh tried to get notification, but the House of Commons had not the nerve. Would you like me to try and get notification made compulsory in the Northern Parliament through our Council? Why not take the bold course?'[79] Unfortunately the response from Dr Rankin, or any further discussion of this issue, does not appear in the archives, but a further letter to Dr Rankin in July 1937 from Dr Thomson illustrates that he had, by that time, decided against compulsory notification: 'compulsory notification, I am afraid, is no good, as it would simply drive the disease underground'.[80] However, rising levels of VD during the Second World War brought about a reversal of Charles Thomson's thinking and, by 1943, he was again arguing for the need to make the diseases notifiable.[81]

The progression of Charles Thomson's views toward VD publicity and education in the interwar years shows much less vacillation than his attitude to compulsory notification. In 1929, and new to the role of MSOH, Thomson described the treatment of VD in moral rather than religious terms and explained that the 'Public Health Department is at one with the theologians' in believing that people needed to 'keep a grip of themselves'.[82] By 1932, however, he had begun to emphasise the need for education and information about VD, as he stressed to the Public Health Committee in this year: 'let us educate, educate, educate at health weeks and in any and every way we can. "Keeping it in the dark" is just like trying to bridge over a festering sore with new skin'.[83] He reinforced these sentiments in an open letter to parents and adults in Belfast, prior to the showing of *Damaged Lives* in September 1934, 'it is a grave matter to allow the rising generation to grow up ignorant of the facts of life, this may result … in much human misery and suffering'.[84] Similarly, discussing the dangers of ignoring the facts about VD in 1937, he asserted that 'it is no use for other people to try and affirm that such things as VD do not exist in Belfast. This is a "hush hush" puritanical attitude and it is so much exalted bunkum! We must get on with publicity and education.'[85]

Although Dr Thomson had consistently argued for increased publicity, propaganda and education, and championed the erection of notices concerning VD in public toilets, very little was actually achieved in the interwar years. He summed up the Belfast Corporation's lack of coherent publicity and propaganda policies when he wrote, in 1942, 'the farthest we have marched has been to employ (some years ago) an expert to lecture males in factories etc; to assent to the exhibition of the film *"Damaged Lives"*; to put up advice bills in certain public places; and to appeal to the leaders of the Churches to warn the young'.[86] It would not be until external circumstances forced the Belfast Corporation to take action to try to prevent the spread of VD during the Second World War that there would be any real advances with propaganda and publicity.[87]

Conclusion

Charles Thomson's correspondence and the changes that took place in his views towards compulsory notification of VD, and with regard to publicity and education, allow an insight into the contemporary debates, or perhaps lack thereof, about VD. The interwar years in Northern Ireland did not see any radical advances in attitudes towards VD or in attempts to prevent the diseases. Unlike other parts of the United Kingdom, there was no widespread medical or public discussion of the issues and, while some special educational events were organised, these were occasional rather than the norm.[88] VD prevention and treatment does not seem to have been high on the agenda of the public health authorities and there appears to have been a marked reluctance to devote much time to an issue that was morally contentious.

The problems associated with establishing a new state that lacked the support of one-third of the population also contributed to the situation. The authorities, both at a local and governmental level, were unwilling to cause unnecessary confrontation and the focus of their attentions was more often on the turbulent political situation, rather than on matters of health. This allowed some local authorities to postpone the establishment of treatment centres in the knowledge that little would be done to force their hand. Similarly, the lack of any strong voices emerging from the medical profession to champion the issue of VD treatment and its prevention ensured that it was not until the crisis created by the Second World War that sustained attention was focused on the subject.

Notes

1   *Royal Commission on Venereal Disease (RCVD), Final Report of the Commissioners,* H.C. 1916 (Cd. 8189) XVI, pp. 84–7.
2   While it is recognised that Northern Ireland did not come into existence until 1921, 'Northern Ireland' is used in reference to the whole period

under discussion and refers to the six counties which came to make up Northern Ireland. The Republic of Ireland was known as the Irish Free State from 1922 until 1937 when it became Éire and then a Republic in 1948.

3 Roger Davidson, *Dangerous Liaisons: A Social History of Venereal Disease in Twentieth-Century Scotland* (Amsterdam, 2000), p. 50; David Evans, '"Tackling the hideous scourge": The creation of Venereal Disease treatment centres in early twentieth-century Britain', *Social History of Medicine*, 5 (1992), p. 424.

4 Susannah Riordan, 'Venereal Disease in the Irish Free State: The politics of public health', *Irish Historical Studies*, 35 (2007), p. 347.

5 For more on the situation in Northern Ireland see, Leanne McCormick, *Regulating Sexuality: Women in Twentieth-Century Northern Ireland* (Manchester, 2009), chapter 4.

6 See for example, Elizabeth Malcolm, '"Troops of largely diseased women": VD, the Contagious Diseases Acts and moral policing in late nineteenth-century Ireland', *Irish Economic and Social History*, 26 (1999), pp. 1–14; Sandra L. McAvoy, 'The regulation of sexuality in the Irish Free State, 1929–1935', in Elizabeth Malcolm and Greta Jones (eds), *Medicine, Disease and the State in Ireland, 1650–1940* (Cork, 1999), pp. 253–66; Philip Howell, 'Venereal Disease and the politics of prostitution in the Irish Free State', *Irish Historical Studies*, 33 (2003), pp. 320–41; Riordan, 'Venereal Disease in the Irish Free State'; Maria Luddy, *Prostitution and Irish Society, 1800–1940* (Cambridge, 2007), chapters 4 and 5.

7 'Londonderry' and 'Derry' both refer to the same city and county and are used in the same way as they appear in the contemporary source material.

8 Tyrone County Council Minutes, 13 February 1919, LA6/2GA/3, Public Record Office of Northern Ireland (hereafter PRONI).

9 Fermanagh County Council Minutes, 27 February 1919, 14 July 1919, LA4/2GA/2, PRONI.

10 Public Health Committee Minutes, Belfast Corporation, 2 December 1920, LA7/9AA/15, PRONI.

11 Fermanagh County Council Minutes, 27 April 1923, LA4/2GA/2, PRONI.

12 *Derry Journal*, 25 April 1919.

13 *Derry Journal*, 16 December 1918.

14 Ibid.

15 Ibid.

16 Ibid.

17 Peter Baldwin, *Contagion and the State in Europe, 1830–1930* (Cambridge, 1999), p. 494.

18 *Derry Journal*, 25 April 1919.

19 County Londonderry VD Committee Minutes, 30 March 1918 and 26 April 1919, LA5/9AK/1, PRONI.

20 County Londonderry VD Committee, Minutes, 1942–1946, LA5/9AK/2, PRONI.

21 List of Treatment Centres for the Treatment of VD, 8 December 1922, Correspondence Concerning the Frequency, Provision for and Treatment of Venereal Diseases, 1922–42, 'B' Files of Ministry of Home Affairs, CAB9B/23/1, PRONI.

22 Boards of Guardians and workhouses existed until 1948 in Northern Ireland, when the National Health Service was introduced. This is in contrast to southern Ireland, where the Boards of Guardians were abolished following

Independence and were replaced by County Boards of Health or Boards of Public Assistance, while workhouses became either County Homes or District Fever Hospitals. In other parts of Britain, workhouses closed throughout the 1920s and 1930s.

23 Davidson, *Dangerous Liaisons*, p. 49.
24 Public Health Committee Minutes, Belfast Corporation, 3 June 1919, 26 August 1919, 23 September 1919, LA7/9AA/13, PRONI.
25 Report of Public Health Committee Deputation, 12 December 1922, LA/7/9BB/15, PRONI.
26 Ibid.
27 Public Health Committee Minutes, Belfast Corporation, 13 April 1926, LA7/9AA/15, PRONI.
28 D. Fletcher to R.G. Gearle, 31 May 1923; R.G. Gearle to D. Fletcher, 1 June 1923, LA/7/9BB/15, PRONI.
29 D. Fletcher to R.G. Gearle, 31 May 1923, LA/7/9BB/15, PRONI.
30 R.G. Gearle to D. Fletcher, 1 June 1923, LA/7/9BB/15, PRONI.
31 Davidson, *Dangerous Liaisons*, p. 50.
32 Public Health Committee Minutes, Belfast Corporation, 15 November 1932, LA/7/9AA/18, PRONI.
33 Ibid.
34 Mary Daly, *A Social and Economic History of Ireland Since 1800* (Dublin, 1981), p. 206.
35 D.S. Johnson, 'The Northern Ireland economy, 1914–1939', in Liam Kennedy and Philip Ollerenshaw (eds), *An Economic History of Ulster, 1820–1939* (Manchester, 1985), p. 213.
36 Greta Jones, *"Captain of All These Men of Death": The History of Tuberculosis in Nineteenth and Twentieth-Century Ireland* (Amsterdam and New York, 2001), pp. 178–9.
37 Prof. Lindsay was Professor of Medicine at Queen's University, Belfast, and had also been involved in the Belfast Eugenics society. There were close links between the National Council for Combating Venereal Disease (NCCVD) and the eugenics movement. For more on the Belfast Eugenics Society see Greta Jones, 'Eugenics in Ireland: The Belfast Eugenics Society, 1911–1915', *Irish Historical Studies*, 28 (1992), pp. 81–95.
38 NCCVD Minutes, 13 May 1918, LA/7/9BB/12, PRONI.
39 Ibid.
40 Public Health Committee Minutes, Belfast Corporation, 13 January 1920, LA7/9AA/13, PRONI.
41 Public Health Committee Minutes, Belfast Corporation, 14 September 1920, LA7/9AA/15, PRONI.
42 Correspondence of Dr Bailie and Mrs Neville-Rolfe, 1919–1921, LA7/9BB/20, PRONI.
43 Mrs Neville-Rolfe to Dr Bailie, 5 April 1921, LA7/9BB/20, PRONI.
44 Ibid.
45 Dr Bailie to Mrs Neville-Rolfe, 19 April 1921, LA7/9BB/20, PRONI.
46 Report submitted Belfast Public Health Committee, 20 September 1920, LA/7/9BB/21, PRONI.
47 Dr Bailie to Mrs Neville-Rolfe, 19 April 1921, LA7/9BB/20, PRONI.
48 Note on VD Propaganda, VD Returns of Cases in 1938 and 1943, Department of Health, B 135/12, National Archives of Ireland (hereafter NAI).

49 Davidson, *Dangerous Liaisons*, pp. 135–42.
50 Ibid., p. 138.
51 Extract from Minutes of Police Committee, 26 June 1919, LA/7/9BB/19 PRONI.
52 Ibid.
53 Extract from Minutes of Public Health Committee, 27 June 1933, LA7/9BB/21, PRONI.
54 Charles Thomson to Chairman and Members of the Improvement Committee, 9 August 1933, LA7/9BB/21, PRONI.
55 Public Health Committee Minutes, Belfast Corporation, 9 January 1934, LA7/9AA/18, PRONI.
56 Charles Thomson to Dr Rankin, 8 July 1937, LA7/9BB/19, PRONI.
57 Public Health Committee Minutes, Belfast Corporation, 12 October 1937, 31 January 1939, LA7/9AA/19, PRONI.
58 Public Health Committee Minutes, 30 June 1942, Belfast Corporation, LA7/9AA/20, PRONI.
59 MSOH for Belfast, *Annual Report of Health*, 1928–1945.
60 Mrs Neville-Rolfe to Charles Thomson, 23 October 1933, LA7/9BB/20, PRONI.
61 For more on the wider context of VD propaganda films see, Annette Kuhn, *Cinema, Censorship and Sexuality, 1909–1925* (London, 1988).
62 Charles Thomson to Mrs Neville-Rolfe, 25 October 1933, LA7/9BB/20, PRONI.
63 Mrs Neville-Rolfe to Charles Thomson, 5 April 1934, LA7/9BB/20, PRONI.
64 Charles Thomson to Mrs Neville-Rolfe, 6 April 1934, LA7/9BB/20, PRONI.
65 *Belfast Telegraph*, 29 September 1934.
66 Rev Doherty to Charles Thomson, 21 September 1934; Rabbi Shachter to Charles Thomson, 23 September 1934, VD Miscellaneous Correspondence, LA7/9BB/19, PRONI.
67 *Belfast Telegraph*, 25 September 1934.
68 *Belfast Telegraph*, 1 October 1934.
69 Ibid.
70 Ibid.
71 'Obituaries', *Journal of the Royal Society for Public Health*, 68 (1948), p. 396.
72 Ibid.
73 Susan Lemar, '"The liberty to spread disaster": Campaigning for compulsion in the control of Venereal Diseases in Edinburgh in the 1920s', *Social History of Medicine*, 19 (2006), pp. 7–10.
74 *Transactions of the Ulster Medical Society, 1917–1929*, becoming the *Ulster Medical Journal*, 1929–1945.
75 Dr Charles Thomson to Dr Rankin, 18 February 1935, LA7/9BB/19, PRONI.
76 Ibid.
77 Ibid.
78 Ibid.
79 Ibid.
80 Charles Thomson to Dr Rankin, 8 July 1937, LA7/9BB/19, PRONI.
81 MSOH for Belfast, *Annual Report of Health*, 1942–43, LA/7/9DA/28, PRONI.
82 Charles Thomson, MSOH Belfast, *The Belfast Book, 1929* (Belfast, 1929), p. 96.
83 Charles Thomson to Public Health Committee, Belfast Corporation, 30 September 1932, LA/7/9BB/19, PRONI.

84 Charles Thomson, Open Letter to the Parents and Adults of Belfast, September 1934, LA7/9BB/19, PRONI.
85 Ibid.
86 Charles Thomson to the Public Health Committee, Belfast Corporation, 6 October 1942, LA7/9BB/21, PRONI.
87 McCormick, *Regulating Sexuality*, chapters 4 and 5.
88 Davidson, *Dangerous Liaisons*, chapter 8; Lemar, 'The liberty to spread disaster'.

10
Moral Prescription: The Irish Medical Profession, the Roman Catholic Church and the Prohibition of Birth Control in Twentieth-century Ireland[1]

Lindsey Earner-Byrne

> the Catholic influence is urgent because medicine is being made more and more a vehicle of attack on the Church e.g. Birth control, Sterilisation of Unfit, Therapeutic Abortion, Psychoanalysis etc.[2]

The history of medicine in Ireland has been complicated by political and religious tensions. Medical institutions sought to make their religious affiliation overt, and often exclusive, while medical appointments were regularly subject to political or religious interference and manipulation.[3] This became more explicit following the foundation of the Irish Free State in 1922, when many Roman Catholic members of the medical profession regarded this political development as an opportunity to ensure that medicine in Ireland become a Catholic stronghold against the perceived machinations of masonry.[4] A solid and successful alliance was cultivated between members of the medical profession and members of the Irish Catholic hierarchy. An important part of this medico-religious alliance was the safe-guarding of an agreed moral stance on certain potential health issues such as birth control and maternal education. The strategies devised to control and regulate these issues according to a Roman Catholic moral view point were crucial in a wider more serious campaign to exclude 'a non-Catholic' influence on the development of medicine in Ireland.[5]

Birth control was prohibited in Ireland from 1935 until 1979.[6] Roman Catholic members of the medical profession were pivotal in shaping the birth control debate in Ireland as a moral rather than a medical issue.

This chapter examines the role of religion, medical lobby groups and international politics in establishing a medico-moral stance on birth control in Ireland, a stance adopted in spite of Irish medical research from the 1930s that highlighted the dangers of repeated pregnancy, and pregnancy for women with certain health conditions.[7] It will also analyse how the issues of birth control and maternal health copperfastened the emerging alliance between Catholic members of the medical profession and the Catholic hierarchy during the first three decades of Irish independence.

For many Catholic lay organisations in Ireland, the foundation of the Irish Free State was regarded as an opportunity to lobby the government to have moral codes enshrined in legislation and, later, in the 1937 constitution. As one regular contributor to the Catholic debate on how the new state should respond to moral issues noted: 'There is now every probability of fixing the legal standard of morality in true consonance with the ideals set before them by the teaching of the Catholic Church.'[8] As well as the Catholic hierarchy, a plethora of groups such as the Catholic Truth Society,[9] the Knights of Saint Columbanus,[10] the Irish Guild of Saint Luke, SS Cosmas and Damian,[11] and the Irish Guild of Catholic nurses[12] were dedicated to this end. For those involved in 'Catholic Action',[13] an essential part of the project of Irish independence was to catholicise the nation, making it a moral bastion in a world deemed to be losing its battle with materialism and immorality.[14] Catholic nurses were regularly called to arms by the Irish Guild of Catholic Nurses to support 'united Catholic Action! How forcibly could Catholic Nurses by means of it make themselves felt in the fight against immoral teaching and practice! (sic)'[15] The Irish Guild of Saint Luke, SS Damian and Cosmas, and the Knights of Saint Columbanus, which shared personnel,[16] were Catholic Action groups particularly prominent in the anti-birth control campaign and in the campaign against a free maternity service in the late 1940s and early 1950s.[17]

Birth control was a potentially fraught moral question as it crossed the line between health and morality. It was therefore essential for those wishing to see a total ban that the subject be seen in absolute terms. Thus, discussion on the issue of birth control was consciously avoided in relation to maternal health in Ireland. Instead, contraception was framed not as a health issue but as a moral one subject to censorship. The Irish Guild of Catholic Nurses, for example, regularly instructed nurses on how to deal with the issue professionally:

> She must know that procured abortion and artificial prevention of conception are, alike, morally wrong. She need not review all the

arguments by which these prohibitions have been arrived at; but she must know that these practices are prohibited and that she may not, for any reason whatever, advise patients to resort to them ... Similarly, she must understand that knowledge gained by her through the exercise of her Profession must be kept a secret and that to do this is a serious obligation.[18]

The Irish Catholic nurse was urged to help combat the modern tendency in her patients to 'complain of the burdens that come with motherhood', as it was 'manifest that the spread of this false gospel prepared the way for unspeakable sin ... a great good may be done by the Nurse, when opportunity offers, in putting forward the Christian view of the privileges of Motherhood'.[19] She was advised not to enter into a discussion regarding birth control with her patients but to merely state: 'I am a Catholic; I know the teaching of the Church on these subjects, they cannot be done without grave sin and I refuse to say more.'[20] The Catholic nurse was a moral guardian of true motherhood stymieing any case made for contraception and censoring information to patients.

Many health workers, including members of the Protestant community like Kathleen Lynn,[21] believed that to consider birth control as part of a maternal health programme would, by association, tarnish the good work of pre-natal clinics and mother and baby clubs. This valuable, life-saving work had to be above all moral suspicion in order to gain the social and religious support so crucial to its success. The issue of 'voluntary motherhood' was further complicated by its association with eugenic principles,[22] as some of the more vocal champions of contraception framed their argument in eugenic terms.[23] Eugenic principles were contrary to Catholic doctrine, so this association made it difficult for any non-eugenic advocates of birth control to build their case on health grounds. The Archbishop of Dublin, Dr Edward Byrne,[24] objected to the amalgamation of St Ultan's Children's Hospital (largely under Protestant control) and the National Children's Hospital at Harcourt Street in Dublin on 'religious principles solely'. He believed that the risk of the children and their parents being exposed to 'naturalistic and wrong teachings on sex instruction or adolescent problems is a powerful argument for retaining the custody of children in Catholic hands'. Highlighting the comments of Prof. Moorhead of Trinity College Dublin on the sterilisation of the unfit, he concluded that an amalgamated hospital would provide the potential for the promotion of 'contraceptive practices'.[25]

Considerable pressure from Catholic lobby groups and the Catholic hierarchy was brought to bear on the Irish government to introduce an uncompromising ban on contraception in Ireland.[26] In February 1926 the Minister for Justice established a committee to consider the need on moral grounds for censorship and prohibition of certain types of printed matter.[27] The committee called on a range of interest groups and witnesses[28] and questioned them on issues regarding immoral literature, including 'magazines published outside Ireland, some of a pseudo-medical character, in which a propaganda in favour of the use of contra-ceptives is carried on'.[29] The committee decried the fact that the law in the Free State regarding the advocacy of the 'limitation of families by unnatural prevention of conception' was 'laxer than in the Dominions, or in the United States, or in many of the Continental European countries'.[30] It noted that in 24 other nations it was a crime to publish or advertise contraceptive information.[31] Other European countries were imposing bans on birth control and abortion; for example, France did so after World War One.[32] However, the Western European imperative was largely a pro-natalist desire to halt the decline in population, rather than any official conviction that birth control was immoral.[33] The committee noted the division of opinion in Great Britain and confidently concluded that no 'similar division of opinion exists in the Saorstát, and that the sentiment expressed by the legislation of America, the Dominions, France, and elsewhere preponderates here'.[34] The committee acknowledged that a few witnesses believed it would be 'an unwanted interference with individual liberty to prohibit the practices', but all 'recognised the dangers attendant on the propaganda as at present conducted'.[35] They recommended that: 'The sale and circulation, except to authorised persons, of books, magazines and pamphlets that advocate the unnatural prevention of conception should be made illegal, and be punishable by adequate penalties.'[36]

The Censorship of Publications Act of 1929, which banned the sale of literature that advocated the use of birth control, was based on the *Report of the Committee on Evil Literature* published in December 1926. While the 1929 Act did not ban the sale or free distribution of contraceptives, it was extremely important because it was information, rather than actual devices, that would have been accessible to the vast majority of people.[37] Furthermore, as the case for birth control in exceptional circumstances regarding maternal health was gaining ground elsewhere in Europe, access to information would have been crucial for this debate to enter the public domain in Ireland. The timing of this ban was interesting as it marked an increasing divergence between Ireland and England with regard to morality and maternal health. Chrystel Hug notes that in the same year as Ireland's Censorship of Publications Act, the Infant Life (Preservation) Act was passed in England, which legalised abortion in circumstances where this would save the mother's life.[38] Perhaps of even greater significance was the fact that in 1930 the British Ministry of Health permitted birth control instruction in maternity welfare clinics 'in cases where further pregnancy would be detrimental to health'.[39]

The deviation between the Roman Catholic and the Protestant view-points on contraception and maternal health was manifest by the early 1930s. In 1930, at the Lambeth conference in England, the Anglican hierarchy, including the Church of Ireland bishops, accepted that birth control was morally justifiable in certain circumstances.[40] In Ireland, the Anglican Bishop of Ossory, Dr Day, explained that there was moral justification in certain 'hard cases'.[41] The decision regarding contraception made at the Lambeth conference was also used by members of the Catholic hierarchy to argue against the appointment of a Protestant as county librarian in Mayo in 1930.[42] The Roman Catholic position was clarified in the same year in the opposite direction with the issuing of the papal encyclical *Casti Connubii*, 'On Christian Marriage'. The encyclical, which outlined the purpose of conjugal relations, addressed the issue of 'hard cases', condemning the argument that a mother's failing health could justify the use of birth control as 'shameful', 'false and exaggerated'. It explained the Catholic position thus:

> Who is not filled with the greatest admiration when he sees a mother risking her life with heroic fortitude, that she may preserve the life of the offspring which she has conceived? God alone ... can reward her for the fulfilment of the office allotted to her by nature, and will assuredly repay her in a measure full of overflowing.[43]

This encyclical was seen in Ireland as validation of the majority opinion and prompted a renewal of the campaign by Catholic lay organisations to close all the legal loopholes in the legislation prohibiting birth control. An essential part of this campaign was the separation of birth control from the health debate. It was in the moral arena that the issue of birth control would be decided and dispatched in Ireland, as Sandra McAvoy has indicated: 'the women's health issue was perceived as subordinate to the public morality question'.[44] The fact that individual women and families would consequently have to carry the burden of such piety, on occasion resulting in the death or physical impairment of the woman, was a price that Catholic campaigners accepted in the interests of preserving a clean national moral conscience.

The Catholic medical ethical view point was clear and uncompromising, and allowed for no 'hard cases'. When a mother's life was potentially endangered by pregnancy, the only ethically acceptable medical course of action was to outline the bald facts to the married couple (for it was never even contemplated that such as situation could arise outside wedlock). Rev. Vincent McNabb, O.P. explained:

> No doctor has the right to say to a married couple: 'You *ought* not to have any children at all; or if at all, then, only after a long interval.' All that may be said is a bare statement of the medical fact, in such words as: 'In my opinion, if you have another pregnancy at any time, or soon, you will die, or be ill, or risk the life of the child etc., etc.'[45]

In 1933, the annual report of the Rotunda hospital provided one damning example of how such moral judgement could have potentially deadly consequences. A 25 year old woman was admitted after swallowing a large amount of abortifacients. After three weeks of procrastination, hyperglycaemia was diagnosed, she was given a minor section, and she died. The Master, Bethel Solomons, candidly admitted that 'This woman might have been saved by an earlier minor section, but her desire for abortion rather cloaked the gravity of her case and probably warped judgement.'[46] The moral agenda was surmountable for some, though, as there is evidence that some middle class women could and did access birth control advice and appliances via sympathetic doctors, while others wrote in desperation to Marie Stopes looking for help and advice.[47] The case of nurse Cadden[48] also revealed that there was a market for the more drastic solution to unwanted or dangerous pregnancies, though sadly for some the solution proved as deadly as the pregnancy might have done.[49]

The issue of birth control and maternal health helped to secure the alliance between the Roman Catholic members of the medical profession and the Roman Catholic hierarchy. Invoking the papal encyclical *Casti Connubii*, the Irish Catholic hierarchy approached the Cumann na nGaedheal government regarding the appointment of Protestant doctors to the state dispensary service.[50] The Archbishop of Tuam, Rev. Gilmartin, raised the issue initially in his Lenten pastoral of 1931:

> In the light of the Papal Encyclical there is revealed a deep cleavage between the Christian idea of marriage and the views of some non-Catholics on this most important subject. Hence a Catholic people have a right that medical practitioners to whom they must have recourse should have been trained in schools where Catholic prin-

ciples about childbirth and kindred subjects are taught, and whose views on the errors and vices condemned by the Pope would be above all suspicion.[51]

This objection to doctors not trained in 'schools where Catholic principles' were imparted was aimed at Trinity College Dublin.[52] Dispensary jobs were highly prized in Ireland for the security they offered and there were frequent and unabashed declarations that such jobs, which involved administering to the largely Catholic poor, should only go to 'safe' Catholic hands.[53] The lack of choice of doctor in the dispensary system was the main reason for objecting to the appointment of any doctors not educated in Catholic medical ethics.[54] Gilmartin's argument against the appointment of Protestant doctors to the dispensary service could have been easily nullified by the simple observation that Protestants were, as all citizens of the state, bound by the law and the law prohibited the 'advocacy of birth control'.[55] The law was surely all the protection Catholic citizens needed. Nonetheless, the Irish Free State government assured Gilmartin that all dispensary doctors would be asked their opinion on contraception before appointment.[56]

Although the Catholic hierarchy was easily mollified on this occasion, the incident provided a preview of future controversy regarding morality and the role of the state in health policy. Gilmartin's fears that poor Catholics would be exposed to 'non-Catholic' principles about childbirth must be understood in the context of the 1930 Lambeth conference and the papal encyclical *Casti Connubii*. It was clear that the international consensus was solidifying on the right to birth control information for those whose health was endangered by future conception. The Catholic doctors' organisation, the Irish Guild of SS Luke, Cosmas and Damian, was critical in providing the anti-contraception campaign in Ireland with a medical justification, and in stressing the need for legislation in accordance with Catholic medical ethics. In June 1932, citing *Casti Connubii*, Dr J. Stafford Johnson, Master of the Irish Guild of SS Luke, Cosmas and Damian, lobbied the League of Nations to change the wording of its most recent report on maternal welfare. The report, entitled *Maternal Welfare and the Hygiene of Infants and Children*, had a passage regarding the distribution of 'anti-conception' advice to mothers suffering from heart disease, tuberculosis or nephritis. The passage to which Stafford Johnson took exception specified that it was not sufficient to tell the mother not to become pregnant, she must also be informed how to avoid pregnancy, if not by her doctor then at a healthcare centre. For Stafford Johnson and his Guild the construction of a clear line between morality and health was essential for Catholic doctors. It was not a question of compassion, it was a question of a rigid and absolutist understanding of morality and the indispensable role of conscience in the provision of health-care. Stafford Johnson explained in his letter to the League of Nations that a Catholic doctor's conscience would never allow him to advocate birth control, irrespective of the claims of maternal health. The issue was clear: contraception was a moral issue not a medical one.[57]

Stafford Johnson had been active throughout the 1930s informing Archbishop Byrne on the denominational landscape of Irish medicine and, in particular, attempting to solidify the influence of Catholics in the reoganisation of the hospital system undertaken by the Hospital Commission.[58] Stafford Johnson made his case to the Archbishop on the basis that the Catholic influence was necessary to ensure the supremacy of Catholic medical ethics, particularly in relation to birth control, sterilisation and abortion.[59] It was Stafford Johnson who contacted the Irish government regarding the League of Nations maternal health booklet.[60] He directed the Department of External Affairs' attention to the offending passage in the League's booklet and noted that birth control was 'repugnant to the feelings of the vast majority of the citizens of Saorstát Éireann'.[61] He pointed to the 1929 Censorship of Publications Act as evidence of national sentiment and he urged the department to take that as a mandate to challenge the League's position and, in so doing, speak up for all Catholic peoples of Europe. The government would become well acquainted with the Guild of SS Luke, Cosmas and Damian: one of its founding members, Dr Edward McPolin, would play an important role in the medical campaign against the introduction of a free mother and child scheme in the late 1940s and early 1950s.[62]

Prompted by Stafford Johnson's urgings the Minister for Local Government and Public Health, Séan T. O'Kelly, concurred that the Irish government's objections should be made on moral grounds, despite the fact that

> Eminent authorities in the fields of gynaecology and obstetrics have borne evidence to the ill effects of the use of contraceptives. Permanent sterility and many neuroses and illnesses, are attributable to this practice [birth control], and some authorities consider that all contraceptives are inimical to the health of both husband and wife.[63]

The Irish government was determined that contraception be viewed as a moral rather than a medical issue, therefore any medical case, however dubious, against birth control was sidelined in deference to the moral argument. O'Kelly also believed that the Irish people would not tolerate an association between birth control and maternal health provision. He noted that

> the practice of contraception is contrary to Catholic doctrine and is abhorrent to the people of Saorstát Éireann and that the association of such teaching with arrangements for maternal welfare would be calculated to bring the health centres into disrepute and to neutralise the efforts that are being made by the State to reduce maternal mortality and morbidity.[64]

O'Kelly had anticipated one of the arguments that would be used against increased state involvement in the sacrosanct arena of maternal welfare. Thus for him, a clear stance that birth control was unacceptable on moral grounds was necessary to avoid any suspicion being cast on state initiatives to reduce maternal mortality. Ironically, this meant denying the legitimacy of the case that birth control might, in certain cases, have saved maternal lives.

The League of Nations was also aware that moral objections to birth control might endanger the progress of wider maternal health issues. The League reassured the Irish government that it was never the intention to advocate the 'indiscriminate use of contraception'; rather it sought to acknowledge the realities of maternal health by ensuring that those whose lives were endangered by pregnancy were 'not called upon to sacrifice their lives'. It argued that it had left the method of birth control vague so as not to 'cause offence to any government or religious body'.[65] However, the Catholic stance on birth control had accepted exactly that possibility and called on mothers to embrace their martyrdom, giving rise to a distinct line of divergence between Protestant and Catholic positions on birth control after the issuance of *Casti Connubii*. The Secretary of the Department of External Affairs, Joseph Walshe, a devout Catholic, suggested wording that was identical to the advice given by Dr McNabb to Catholic doctors in 1926 and that was the policy stance of the Irish Guild of SS Luke, Cosmas and Damian:

> Nevertheless it may be necessary to avoid pregnancy on account of the mother's own health. But it is not sufficient merely to tell a married woman suffering from tuberculosis or heart disease or nephritis that she should not become pregnant. The dangers, which

might arise from pregnancy in such cases, should be fully explained by a doctor to the husband and wife.[66]

The Irish government therefore stopped short of advocating maternal martyrdom, but it nonetheless refused to make any moral exception to its categorical stance on the issue of birth control. A married couple was only entitled to information that would lead to a morally acceptable choice. A husband and wife, therefore, were entitled to the information regarding the risks of pregnancy and the potential consequences, but not to any information on how to avoid pregnancy. As one doctor, who wrote to the Department of Local Government and Public Health regarding the issue, explained, it was a doctor's duty to inform a woman of the risk to her life, 'but it should be left to the patient as to what steps they would take to avoid the possibility of becoming pregnant'.[67]

As a result of Irish lobbying the League of Nations reworded the passage, entering the following caveat: 'It is necessary further that the steps to be taken to avoid this should be explained to the husband and wife by a doctor, either privately or at a health centre, due account being taken of the individual's religious beliefs and moral principles, as well as of national legislation.'[68] Despite the fact that the Taoiseach and Minister for External Affairs, Eamon de Valera regarded this inclusion as a 'success' in the face of 'very considerable opposition',[69] publicly the Irish Free State responded that the section was 'entirely contrary to Catholic teaching'.[70] Only Italy supported the Irish position fully, both Spain and Britain objected.[71]

Mark Finnane pointed out that the Carrigan Committee, established to review the Criminal Law Amendment Acts of 1880 and 1885, also dealt with the issue of contraception in late 1932 at its final meeting. At the same time as the Department of External Affairs was arguing against birth control for 'hard cases', the Carrigan Committee, under the purview of the Minister for Justice, James Geoghegan (a member of the Catholic Truth Society)[72] recommended that, while there should be a general prohibition on contraceptives, 'qualified medical practitioners might have the power to prescribe and supply such appliances'.[73] The committee's recommendation, which hinted at a medical case for contraception and made the doctor the pivotal arbiter between health and morality, was not followed up in subsequent legislation. Despite the appearance of this clause in Article 16 of the original draft of the Criminal Amendment Bill, it was removed from the final draft of the Bill at the behest of the Séan T. O' Kelly.[74]

In view of the Irish government's campaign to have the League of Nations' maternal health booklet altered, it is not surprising that O'Kelly moved to ensure Irish legislation unequivocally banned contraception regardless of health issues. The compromise wording that the League settled on included the necessity to respect 'national legislation' when advising a mother of the risks to her health and any future preventative measures she might take. The Irish state would have looked rather foolish if its own legislation had facilitated the very situation it had fought to prevent being imposed on it by the League of Nations. The Fianna Fáil government introduced the Criminal Law Amendment Act of 1935, which addressed most of the loopholes in the previous legislation of 1929 regarding the sale of contraceptives and allowed for no medical exception to the total ban on contraceptive information or appliances. This act effectively instituted an unequivocal ban on contraception, which remained in force until contraceptives were legalised for married couples under the Health (Family Planning) Act, 1979.[75]

The deliberate separation of birth control and maternal health was very significant in Ireland for two major reasons: first, the demographic profile of late marriages and large families and, second, the high rate of tuberculosis.[76] Repeated pregnancies borne by older mothers increased the risks of complications, and tuberculosis was one of the conditions identified as making pregnancy an increased risk to female health. Nevertheless, the *status quo* in relation to birth control was rarely challenged publicly in Ireland. Bethal Solomons, the Jewish Master of the Rotunda maternity hospital between 1926 and 1933, who was not against artificial birth control,[77] only dared to imply his dissent by criticising the impact of regular pregnancy on poor women.[78] Solomons studied the physical and mental cost of multiple pregnancies on Dublin's poorest mothers. His references to 'quality of life' contained implicit criticism of a state and society that left the most vulnerable individuals with few options when their health needs clashed with the state's moral imperatives. He was concerned not merely with those women whose lives were jeopardised by pregnancy, but also with the countless poor women he treated who were debilitated by relentless pregnancy, childbearing and childrearing. He coined the phrase 'dangerous multipara' to describe a mother who had given birth more than seven times. He first used this expression in *The Lancet* in 1934, when he cautioned that 'it is altogether a mistake to suppose that in childbearing practice makes perfect'.[79] He noted that 'from the fifth pregnancy the rate of mortality is over average by an amount which increases steadily and speedily until in women bearing their tenth child or more the mortality is five times as high as for all women bearing children'.[80] In 1933, he cautioned that '[w]omen with decompensated cardiac disease should not marry, and if they do, they should not become pregnant, or they will surely die'.[81]

In 1932, the Department of Local Government and Public Health articulated a similar view declaring that it was 'a fact that pregnancy is a grave complication in the case of a woman suffering from any form of tuberculosis ... tubercular patients should not marry'.[82] This acknowledgment amounted to an admission that the state policy on birth control meant that a large number of women affected by either tuberculosis or heart disease engaged in sexual activity at their peril. This was, however, not information that was given wide public coverage: there was no campaign to warn this vulnerable and identifiable group of women of the risks posed by sexual relations. Revealingly, this departmental memorandum was written in relation to the campaign against the wording of the League of Nations' maternal health handbook; thus, it was a clear acknowledgment of the price of the total ban on contraception that the Irish state wanted and would later enshrine in national legislation.

The health issue was raised in the Seanad debate regarding the 1929 Censorship of Publication Bill and during the Dáil and Seanad debates for the 1935 bill, but only by the few Protestant voices available. Dr Rowlette was a lone voice in the Dáil, and in the medical profession, when he attempted to argue the health and conscience case in 1934. During the Dáil debate on the Criminal Law Amendment Bill in 1934 he presented the case of a 'respectable married woman in ill health that would endanger her life by pregnancy'.[83] He was careful to express agreement with the 'common moral judgement and feeling in this country', which was against the practice of birth control in all circumstances. He pointed out that, while abstinence may be a viable alternative for respectable women, 'there are very considerable practical difficulties in many households ... with narrow accommodation, where a wife and her husband are living on affectionate and proper terms'. He argued that abstinence under such circumstances could lead to 'grave nervous disorders' in the wife, 'infidelity on the part of the husband', and 'marital unhappiness' all round.[84] Rowlette boldly touched on the private grief many couples must have been burdened with when a choice between abstinence or possible death confronted them in the marital bed. Of course, the very private nature of it made it possible to maintain the rigid stance against contraception, for how many of these couples would feel free to speak out publicly about their predicament? If public representatives and members of the medical profession failed to make this a political issue who else could? Even Ireland's women's movement avoided the subject of contraception because it was too divisive for a movement that craved cohesion.

The fact that the Roman Catholic Church made a virtue out of maternal self-sacrifice presented a virtually insurmountable obstacle to the maternal health argument. The Catholic Church advocated that a wife 'must resist the intercourse of a husband who uses a condom, as a virgin must resist a man who attacks her virtue'. According to the advisor of the *Irish Ecclesiastical Record* this advice from the Catholic Church meant that women were supposed to employ 'active, forceful resistance which may be discontinued only in the face of the gravest actual danger – and provided there is no proximate fear of consent to the illicit intercourse'.[85] Thus it was the wife that should function as the ultimate and final defence against the introduction of birth control into the marital bed. Evidently, according to the Roman Catholic Church, female health and safety was second to virtuous morality and the prevention of birth control. Against this unforgiving stance, Dr Rowlette's plea for individual conscience, and for the rights of those people in the state 'who do not adopt the view that the use of contraceptives in such a position [maternal health] is contrary to morality', fell on resolutely deaf ears.[86]

Sean T. O'Kelly's prediction in 1932, during the League campaign, that any association between birth control and maternal health initiatives could be detrimental to the progress of the latter was proved correct in the opening salvo of the mother-and-child controversy. In 1947 the new Minister for Health, Dr James Ryan, published a Public Health Bill, which had had a protracted and complicated incubation,[87] with a separate section entitled 'mother and child'. This section outlined the official commitment to 'make arrangements for safe-guarding the health of women in respect of motherhood and for their education in that respect'.[88] The bill was enacted in August 1947 and followed by a White paper, *Outline of Proposals for the Improvement of the Health Services*, which made it clear that provisions for mothers and children outlined in the 1947 Act were just the start of a long-term programme to provide free medical care to all sections of society.[89] On 7 October 1947 the Bishop of Ferns, Dr James Staunton, in his capacity as the secretary to the Roman Catholic hierarchy's standing committee, sent a statement to the Taoiseach on the new health act.[90] The hierarchy believed that the rights of the family and the Church were violated by the state's intention to take over the education of mothers in respect of motherhood, as it was 'precisely in this sphere of health education, where so many moral questions arise'.[91] Ruth Barrington has concluded that the anxieties apparent in this formal protest were indicative of the deep fears regarding the introduction of birth control in Ireland.[92] While, as Barrington points out, the principle of public authorities advising women in respect of motherhood was already introduced in the Notification of Births Act, 1915 and the Public Health (Medical Treatment of Children) (Ireland) Act, 1919,[93] these acts had not been introduced by an Irish administration and the 1915 Act was permissive, unlike the 1947 Act. The Irish Catholic hierarchy were reacting to the changed climate of 1947, where it was highly conceivable that the Irish state was preparing the ground for the introduction of a level of state medicine; in such a world, the case of maternal health and contraception might have received fresh attention and even resulted in a change in the state's legal position. The National Health Service in Britain and its proposed introduction in Northern Ireland[94] was causing considerable alarm in Irish Catholic and medical circles.[95] The new maternity services proposed in the 1947 Health Act provided for the education of all mothers in respect of health, without allowing for a choice of doctor, and the educative clause was vague enough to cause alarm regarding the possible provision of birth control. This raised the old ghosts of 1931, when Gilmartin had spoken out against the moral threat of Protestant dispensary doctors.

E. McKee claims that what brought the Irish Catholic hierarchy into the campaign against the mother-and-child scheme 'was the successful demonstration [by the Irish Medical Association] of the innate connexion between the survival of private practitioners in public health and the safeguarding of Catholic morality and social teaching'.[96] The most vocal members of the medical profession consciously married their professional concerns regarding increased state power with Catholic Social teaching. Dr James McPolin, a founding member of the Irish Guild of SS Luke, Cosmas and Damian, judiciously employed Catholic doctrine to stir up profound anxieties in medical circles regarding state medicine.[97] In July 1947 McPolin sent extracts from papal encyclicals to the Department of Health as proof that the Health Bill would 'damage family life'.[98] McPolin's arguments formed the basis of the medico-moral opposition to any new maternity scheme that was designed and controlled by the state. McPolin's main contention was that 'the only true object of law and medical activity is the promotion of virtue and any other view of law and medicine will eventually corrupt health as well as virtue'.[99]

For political reasons the mother-and-child scheme was not finally outlined until June, when Fianna Fáil had been replaced by the first inter-party government, with Noel Browne as Minister for Health.[100] The religious and medical objections were swift and amounted to a reiteration of previous concerns outlined in 1947. The Catholic hierarchy regarded the proposed scheme, which sought to provide free medical care to all mothers and children up to 16 years, as 'a ready-made instrument of future totalitarian aggression' with the potential to provide 'birth limitation and abortion'.[101] The Minister for Health did little to assuage such concerns when he claimed that the education of mothers 'may include instruction in sex relationship, chastity and marriage. It also includes, however, such matters as correct diet during pregnancy'.[102] In reality, for both the Catholic hierarchy and the Irish medical profession, the issue of contraception was closely allied to the wider dread of socialised medicine and state control, which was deeply feared by both parties for professional and moral reasons. Browne's scheme failed for a myriad of reasons too complex to go into here; however, the issue of contraception and abortion remained central to all objections and in the compromise scheme, introduced in 1953, there was no question of 'instruction on sex' or contraception being included. The fear of contraception and abortion was symbolic of a lack of trust in any state-controlled maternity service and a conviction that this would lessen the Catholic influence on Ireland's health services. It is no coincidence that after the introduction of the compromise mother-and-child scheme in 1953, McQuaid accused de Valera of a 'latent anticlericalism that fears the influence of the Church and will always seek to eliminate that influence from public life'.[103]

Throughout this period the Irish Catholic hierarchy and members of the Irish medical profession involved in the Catholic Action movement, worked to ensure that health and birth control remained separate issues in Ireland. Many in the medical profession were deeply influenced by Catholic social teaching and used papal teaching to bolster their arguments against any form of state medicine and to promote Catholic control of medicine. Through organisations like the Irish Guild of SS Luke, Cosmas and Damian and the Knights of Saint Columbanus, they rubbed shoulders with other professionals, among them government ministers who nurtured a similar Catholic vision. This was an often secret and very powerful world that operated to ensure Catholic principles and Catholic professionals remained central to developments in the Irish medical scene. Stafford Johnson, for example, was a member of both these organisations and, along with McQuaid, was pivotal in ensuring the Catholic control of medicine in Dublin.[104] Birth control became a central issue in this powerful alliance and a benchmark of the degree to which Catholic principles shaped medical boundaries in Ireland.

Notes

1   Sections of this article are taken from L. Earner-Byrne, *Mother and Child: Maternity and Child Welfare in Dublin, 1922–60* (Manchester, 2007), pp. 39–47.

2   'Guild of St Luke, SS Damian and Cosmas, 'Memorandum: The proposed reorganisation of Irish hospitals', ca. 1935, Archbishop Byrne papers, Hospital Commission Box, DDA. Cited in K. Morrissey, 'An examination of the relationship between the Catholic Church and the medical profession in Ireland in the period 1922–1992, with particular reference to the impact of this relationship on the field of reproductive medicine' (unpublished PhD thesis: University College Dublin, 2004), p. 91.

3   Laurence Geary, *Medicine and Charity in Ireland, 1718–1851* (Dublin, 2004), pp. 157, 159–61; M Ó hÓgartaigh, 'A medical appointment in county Meath', *Ríocht na Midhe: Records of the Meath Archaeological and Historical Society*, 17 (2006), pp. 266–70; Greta Jones notes that 'denominational rivalry', which existed prior to the foundation of the Irish Free State, intensified thereafter. G. Jones, *'Captain of all these Men of Death': The History of Tuberculosis in Nineteenth and Twentieth-Century Ireland* (New York, 2001), p. 134.

4   Noel Browne claimed that medical appointments in voluntary hospitals were often made on religious grounds, with Protestant-dominated hospitals operating through the Masonic Order and Catholic-dominated ones operating through the Reverend Mother or the Knights of Saint Columbanus. N. Browne, *Against the Tide* (Dublin, 1986), p. 146. White notes that there was disquiet among the Protestant community regarding the role played by the Masons in securing hospital appointments. J. White, *Minority Report: The Protestant Community in the Irish Republic* (Dublin 1975), p. 167.

5   The campaign to ensure Catholic control of hospitals in the Free State was largely based on arguments that Catholic medical ethics needed to be safeguarded. 'Guild of St Luke, SS Damian and Cosmas, 'Memorandum: The proposed reorganisation of Irish Hospitals'.

6   Ireland in this article refers to the 26 counties of the Irish Republic.

7   Medical research either emanating from Ireland or current in Irish medical circles indicated that the following conditions posed problems to women's health during pregnancy: tuberculosis, heart disease, nephritis, malnutrition. See, for example, Memo Department of Local Government and Public Health, c. 1932, National Archives of Ireland [hereafter NAI], Dept. Health, B130/59; Bethel Solomons, 'The prevention of maternal mortality and morbidity', *Irish Journal of Medical Science*, 88 (April 1933), p. 175.

8   R.S. Devane, 'The unmarried mother: Some legal remedies', *Irish Ecclesiastical Review*, 23 (January–July 1924), p. 58.

9   Established at the end of the nineteenth century, the Catholic Truth Society played a prominent role in driving Catholic Action in early independent Ireland and was a leading member in the campaign against immoral literature and birth control.

10 Founded in Belfast in 1915 and in Dublin in 1922, the Knights of Columbanus was a secretive organisation that sought to advance the temporal

interests of Catholics in Ireland particularly in relation to professional appointments. It was involved in certain Catholic 'take-overs' of medical initiatives and in one instance of a hospital board. See E. Bolster, *The Knights of Saint Columbanus* (Dublin, 1979).

11   Dr Stafford Johnson wrote to Archbishop Byrne in 1931, to inform him of the establishment of this Roman Catholic organisation of doctors to ensure 'the application of Christian virtue in the practice of their profession and life'. DDA, Byrne papers box one: lay organisations, Stafford Johnson to Archbishop Byrne, 31 November 1931. See also, the inaugural address delivered before the Guild of SS Luke, Cosmas and Damian, Dublin, 21 April 1932, 'The true idea and outlook of a Catholic medical guild' by Stafford Johnson. Reprinted from the *Catholic Medical Guardian* (July 1932).

12   The Irish Guild of Catholic Nurses was established in August 1922 to ensure the supremacy of Catholic morality in Irish nursing. See *Irish Nursing News*, 5:11 (August 1927), p. 619.

13   See for example, Rev. E. Cahill, *The Catholic Social Movement* (Dublin, 1931). For a detailed discussion see, E. Dunne, 'Action and reaction: Catholic lay organisations in Dublin in the 1920s and 1930s', *Archivium Hibernicum*, 48 (1994), pp. 107–18.

14   For a detailed discussion of this, see J.H. Whyte, *Church and State in Modern Ireland, 1923–1970* (Dublin, 1974), pp. 24–61.

15   'Excerpt from the Irish Guild of Catholic Nurses solemn renewal given by Rev. H. Kelly on 30 October, 1927', *Irish Nursing News*, 6:2 (November 1927), p. 36.

16   Stafford Johnson was Supreme Knight of the Knights of Columbanus from 1942 to 1948 and a founding member of the Guild of St Luke, SS Damian and Cosmas.

17   The dispensary system was the principal health service available to the poor in Ireland. It has a complex history dealt with in some detail in this collection by Catherine Cox. It had a reputation for corruption and cronyism particularly in relation to appointments, which were sought for their relative security and entitlements and, after independence, there was a concerted effort to 'clean up' the system. The Local Authorities Officers and Employers Act, 1926, sought to secure a national system run on 'merit alone', which went some way to ending political favouritism and nepotism. However, the dispensary system suffered neglect and virtual breakdown in certain areas throughout the 1930s. See R. Barrington, *Health, Medicine and Politics in Ireland 1900–1970* (Dublin, 1987), pp. 101, 134.

18   'Excerpt from Irish Guild of Catholic Nurses annual triduum given by Rev. J.E. Canavan, 14 September 1928', *Irish Nursing News*, 7:2 (November 1928), p. 28.

19   'Excerpt from the Irish Guild of Catholic Nurses' monthly day of recollection given by the Very Rev. J. Kearney', *Irish Nursing News*, 7:5 (February 1929), p. 76.

20   Excerpt from 'The nurse in the twentieth century': The sixth annual conference held under the auspices of the Irish Guild of Catholic Nurses', *Irish Nursing News*, 8:6 (March 1930), p. 86.

21   Kathleen Lynn founded St Ultan's Children's Hospital. See M. Ó hÓgartaigh, *Kathleen Lynn: Irishwoman, Patriot, Doctor* (Dublin, 2006).

22   Catriona Clear, 'Women in de Valera's Ireland 1932–48', in G. Doherty and D. Keogh (eds), *De Valera's Irelands* (Cork, 2003), p. 113; Ó hÓgartaigh, *Kathleen Lynn*, pp. 96–106.

23   Some of the only arguments put forward in the Dáil and Seanad in favour of limited access to birth control, during the lead up to the 1929 and 1935 censorship acts, were laced with eugenic ideas. See, for example, Sir John Keane's contribution *Seanad Éireann Debates*, vol. 12, col. 67 (11 April 1929).

24   Dr Edward Byrne (1872–1940) served as Archbishop from 1921 until 1940.

25   Statement of His Grace, the Archbishop of Dublin, to members of the Deputation of the Committee of St Ultan's Hospital, 20 December, 1935, Archbishop Byrne papers, Hospital Commission Box, DDA.

26   For a survey of the mounting pressure regarding evil literature (often code for information relating to birth control) see Rev. R.S. Devane, S.J., 'Indecent literature: Some legal remedies', *Irish Ecclesiastical Review*, 25 (January–June 1925), pp. 182–204.

27   *Report of the Committee on Evil Literature* (Dublin, 1926), p. 3.

28   The following organisations were invited to testify: Catholic Truth Society, City of Dublin Young Men's Christian Association, Church of Ireland Young Men's Christian Association, Irish Vigilance Association, Dublin Christian Citizenship Council, Father Devane, Catholic Writers' Guild, Boys' Brigade, Boy Scouts, Superior-General, Christian Brothers, Catholic Headmasters' Association, School Masters' Association, Secretary, Dublin Branch, Irish National Teachers' Organisation, Marian Sodalities of Ireland. The following do not appear on the list of those that actually testified: The School Masters' Association, Church of Ireland Young Men's Christian Association, the Boys' Brigade, Boy Scouts and the City of Dublin Young Men's Association. Other witnesses included representatives from Eason & Son, Ltd., Irish Retail Newsagents, Booksellers and Stationers' Association, Department of Posts and Telegraphs, Superintendent Inspector of Customs and Excise, and the Deputy Commissioner of An Gárda Síochána (the police force of the Irish Free State). See *Report of the Committee on Evil Literature*, pp. 4–5.

29   Ibid., p. 5.

30   Ibid., p. 13.

31   Ibid.

32   Franco's Spain also introduced a prohibition on birth control and information relating to contraception as did Italy's Mussolini. For a full discussion of the background to each country see, E. Ketting, *Contraception in Western Europe: A Current Appraisal* (London, 1990).

33   L. Hollen Lees, 'Safety in numbers: Social welfare legislation and fertility decline in Western Europe', in J.R. Gillis, L.A. Tilly and D. Levine (eds), *The European Experience of Declining Fertility, 1850–1970: The Quiet Revolution* (Oxford, 1992), pp. 310–25. The spectre of 'race suicide' and the case of France was repeatedly brought up in the official debates regarding the ban on literature that advocated birth control. See, for example, *Dáil Éireann Debates*, vol. 26, col. 687 (19 October 1928).

34   *Report of the Committee on Evil Literature*, p. 14

35 Ibid.

36 Ibid., p. 19. They also recommended a ban on 'indecent advertisement' relating to sexual matters, irregularities of menstruation, and prevention of conception, etc.

37 Mary E. Daly also made this point in '"An Irish solution to an Irish problem": The medical profession and Irish family planning', unpublished paper presented to the Centre for the History of Medicine in Ireland, UCD, 19 February 2009.

38 C. Hug, *The Politics of Sexual Morality in Ireland* (New York, 1999), p. 79.

39 Memorandum 153/MCW cited in L. Hoggart, 'The campaign for birth control in Britain in the 1920s', in Anne Digby and John Stewart (eds), *Gender, Health and Welfare* (London, 1996), p. 144; R.A. Soloway, *Demography and Degeneration: Eugenics and the Declining Birthrate in Twentieth-Century Britain* (Durham, NC, 1995), pp. 182–8; for a discussion of the impact of these issues see W.J. O'Donovan, 'Medical opinion and its influence on population', in *Sixth International Congress of Catholic Doctors* (Dublin, 1954), pp. 86–94.

40 This change in sentiment on the part of the Anglican Church was noted in official debates in Ireland as early as 1929. See, for example, *Seanad Éireann Debates*, vol. 12, cols. 67–8 (11 April 1929).

41 *Kilkenny People*, 31 August 1930.

42 P. Walsh, *The Curious Case of the Mayo Librarian* (Cork, 2009), p. 16.

43 *Casti Connubii*, quoted in A. Fremantle, *The Papal Encyclicals in their Historical Context* (New York, 1956), p. 239.

44 Sandra McAvoy, 'Regulation of sexuality in the Irish Free State', in E. Malcolm and G. Jones (eds), *Medicine, Disease and the State in Ireland, 1650–1940* (Cork, 1999), p. 257.

45 Cited in the Catholic Truth Society, *The Problem of Undesirable Printed Matter: Suggested Remedies* (Dublin, 1926), p. 11. [Author's emphasis.]

46 'The annual report of the Rotunda hospital', *Irish Journal of Medical Science*, Ser. 6, no. 92 (August 1933), p. 334.

47 Cormac Ó Gráda, *Ireland: A New Economic History, 1780–1939* (Oxford, 1994), p. 219. See also Bethel Solomons, *One Doctor in his Time* (Dublin, 1952).

48 Mary Anne (Mamie) Cadden served prison sentences for child abandonment and abortion in 1939 and 1945 respectively. In 1957 she was found guilty of the murder of Helen O'Reilly, who died following an abortion administered by Cadden. See S. McAvoy, 'Before Cadden: Abortion in mid twentieth-century Ireland', in D. Keogh, F. O'Shea, and C. Quinlan (eds), *The Lost Decade: Ireland in the 1950s* (Cork, 2004), pp. 147–63; E.L. Deale, *Beyond Responsible Doubt? A Book of Irish Murder Trials* (Dublin, 1990); NAI, Dept. Taoiseach, S16116, 'Mary Anne Cadden: Death sentence, 1956–7'.

49 McAvoy, 'Before Cadden'; P. Conroy, 'Maternity confined – the struggle for fertility control', in P. Kennedy (ed.), *Motherhood in Ireland: Creation in Context* (Cork, 2004), p. 131; F. Kennedy, *Cottage to Crèche: Family Change in Ireland* (Dublin, 2001), p. 39.

50 Dept. Taoiseach, S2547A. NAI.

51 Ibid.

52   Solomons notes that Catholic doctors educated in Trinity College were also subjected to bans in appointments in some Catholic hospitals. Solomons, *One Doctor in His Time*, p. 110.

53   Geary, *Medicine and Charity in Ireland*, pp. 123–51; Ó hÓgartaigh, 'A medical appointment in Meath', pp. 266–7.

54   For a detailed account of Catholic ethics and the practice of medicine in Ireland see, Morrissey, 'An examination of the relationship between the Catholic Church and the medical profession in Ireland in the period 1922–1992.'

55   Censorship and Publication Act, 1929.

56   Memorandum of conversation between the Archbishop of Tuam and Sir Joseph Glynn, 26 February 1931. NAI, Dept. Taoiseach, S2547.

57   Letter from Stafford Johnson to the League of Nations, 6 June 1932, International League of Nations legislation relating to maternity and child welfare, NAI, Dept. Health, B130/59.

58   See Guild of St Luke, SS Damian and Cosmas, 'Memorandum: The proposed Reorganisation of Irish Hospitals' ca. 1935, Archbishop Byrne papers, Hospital Commission Box, DDA. This is a fascinating document which itemises the religious control of each hospital.

59   Ibid.

60   Dept. Health, B130/59. NAI.

61   Ibid.

62   For a detailed explanation of the mother-and-child controversy see E. McKee, 'Church-state relations and the development of Irish health policy: The mother-and-child scheme, 1944–53', *Irish Historical Studies*, 25:98 (November 1986), pp. 159–94.

63   Note signed by ED, 23 August 1932, NAI, Dept. Health, B130/59.

64   Letter from the Department of Local Government and Public Health to the Department of External Affairs, 8 September 1932, NAI, Dept. Health, B130/59.

65   Extract of Coyne's report: 'Confidential statement re section on contraception issued by the League of Nations', c. 1932, NAI, Dept. Health, B130/59.

66   Secretary of the Department of External Affairs, 17 November 1932, NAI, Dept. Health, B130/59.

67   Dr McDonald to the Department of Local Government and Public Health, 28 July 1932, NAI, Dept. Health, B130/59.

68   *Report of the Council on the Work of the Nineteenth Session of the Committee*, 15 October 1932, NAI, Dept. Health, B130/59.

69   Memo from Department of External Affairs, 8 September 1933, NAI, Dept. Health, B130/59.

70   *2nd Committee of the 14th Assembly of the League of Nations*, 28 September 1933, NAI, Dept. Health, B130/59.

71   The German and French recommended a sentence repudiating any intention to advocate contraception. The Polish delegates sympathised with the Irish position but saw difficulties in implementation. See, 'Memo: Work of the Health Organisation', 6 October 1932, NAI, Dept. Health, B130/59.

72   P.J. Dempsey, 'Geoghegan, James', in J. McGuire and J. Quinn (eds), *The Dictionary of Irish Biography*, 4 (Cambridge, 2009), pp. 50–1.

73 Cited in M. Finnane, 'The Carrigan Committee of 1930–1 and the "moral condition of the Saorstát"', *Irish Historical Studies*, 33:128 (November 2001), p. 529.

74 See Finnola Kennedy, 'The suppression of the Carrigan report: A historical perspective on child abuse', *Studies*, 89, 356 (Winter 2000), pp. 354–63.

75 Hug, *The Politics of Sexual Morality in Ireland*, p. 114.

76 R. Geary, 'The future population of Saorstát Éireann and some observations on population statistics', *Journal of the Statistical and Social Inquiry Society of Ireland*, 15 (1935–6), pp. 15–35, p. 20; For a comprehensive analysis of Irish demographic patterns see, M.E. Daly, *The Slow Failure: Population Decline and Independent Ireland, 1920–1973* (Wisconsin, 2006), pp. 3–20, 75–137 and M.E. Daly, 'Marriage, fertility and women's lives in twentieth-century Ireland (c. 1900–c. 1970)', *Women's History Review*, 15:4 (September 2006), pp. 571–85. See also, Jones, *'Captain of all these Men of Death'*.

77 Solomons, *One Doctor in His Time*, pp. 91–2; Bethel's son claimed his father gave contraceptive advice when asked by patients. M. Solomons, *Pro Life? The Irish Question* (Dublin, 1992), p. 16.

78 B. Solomons, 'The dangerous multipara', *The Lancet*, 2 (7 July 1934), pp. 8–11.

79 Ibid.

80 Ibid.

81 Solomons, 'The prevention of maternal morbidity and mortality', *Irish Journal of Medical Science*, Ser. 6, no. 88 (April 1933), p. 175.

82 Memo Department of Local Government and Public Health, c. 1932, NAI, Dept. Health, B130/59.

83 Cited in Kennedy, *Cottage to Crèche*, p. 163.

84 *Dáil Debates*, vol. 53, col. 2019 (1 August 1934).

85 'Notes and queries: The excusing cause for "passivity" in onanism. Counselling the lesser sin of onanism', *Irish Economic Review*, 70 (January–June 1948), p. 245.

86 *Dáil Debates*, vol. 53, col. 2019 (1 August 1934).

87 Earner-Byrne, *Mother and Child*, pp. 120–6.

88 Health Bill, 1947, section 21, p. 12.

89 Barrington, *Health, Medicine and Politics in Ireland*, p. 188.

90 Statement of the Irish hierarchy on the Health Act, 1947, p. 1, DDA, McQuaid papers, AB8/B/XVIII/14. Also in NAI, Dept. Taoiseach, S13444A.

91 Ibid., p. 2.

92 Barrington, *Health, Medicine and Politics in Ireland*, pp. 186–7, 190.

93 For an explanation of both Acts see Barrington, *Health, Medicine and Politics in Ireland*, pp. 76–7, 80.

94 The NHS was introduced to Northern Ireland in July 1948.

95 Barrington, *Health, Medicine and Politics in Ireland*, p. 188; T. Farmar, *Patients, Potions and Physicians: A Social History of Medicine in Ireland* (Dublin, 2004), pp. 170–4.

96 McKee, 'Church-state relations and the development of Irish health policy', p. 185.

97 James McPolin, 'Some aspects of the sociology of the medical profession', *Journal of the Medical Association of Éire* (hereafter *JMAÉ*) 19:110 (August

1946), pp. 118–22, 135–44; James McPolin, 'Doctors and professional secrecy', *JMAÉ*, 23:107 (September 1948), pp. 39–41.

98   McPolin to Department of Health, 9 July 1947, NAI, Dept. Taoiseach, S13444E.

99   McPolin, 'Doctors and professional secrecy', p. 40.

100 For a detailed analysis of the medical position see Earner-Byrne, *Mother and Child*, pp. 130–44.

101 Dr James Staunton, secretary of the hierarchy to the Taoiseach, 10 October 1950, DDA, McQuaid papers, AB8/B/XVIII/15.

102 'Memo of observations of the Minister for Health on various matters relating to the mother-and-child scheme referred to in a letter, dated 10 October 1950, addressed to the Taoiseach by the Most Rev. J. Staunton, DD, Bishop of Ferns, Secretary to the Hierarchy' (undated, c. 12 October 1950), p. 4. Copy in DDA, McQuaid papers, AB8/B/XVIII/15.

103 McQuaid to Levame, 14 April 1956, DDA, McQuaid papers, Nuncio Box.

104 According to David Sheehy, Stafford Johnson was a close personal friend of John Charles McQuaid, Archbishop of Dublin (1941–1972), frequently dining with him at his private residence in Killiney, county Dublin. D. Sheehy, 'Archbishop McQuaid: The Diocesan Administrator', *Doctrine and Life*, 54:3 (March 2003), p. 168.

11
Death and Disease in Independent Ireland, c. 1920–1970: A Research Agenda

Mary E. Daly

Introduction

The analysis of mortality and disease in independent Ireland must be set in the context of Ireland's[1] demographic and socio-economic profile. The Famine of the 1840s was followed by more than a century of population decline, which was caused by the mass emigration of young adults, a practice that was not reversed until the 1960s. Consequently, the proportion of the population who were either elderly or children, both categories with high mortality, was significantly greater than in other countries. Another characteristic, which might have a bearing on female mortality and morbidity, was the persistence well into the twentieth century of high marital fertility, coupled with a late age of marriage, resulting in a high birth rate among women over the age of 35, which can be seen in the data in Table 11.1.

The population of the 26 counties that constituted the independent Irish state was predominantly rural and agrarian until the 1960s. Irish cities and towns outside Ulster were in many respects pre-industrial, or non-industrial places, with a high proportion of men engaged in casual labouring and significant levels of chronic underemployment.[2] Family poverty was not mitigated to any significant extent by the earnings of

Table 11.1 Mean number of children born per 100 women with marriages of 20–24 years' duration

1911	(married 1887–1891)	596
1946	(married 1922–26)	449
1961	(married 1937–41)	416
1971	(married 1947–51)	424

married women or co-resident children; opportunities for paid employment for women and teenagers were significantly fewer than in Britain or Northern Ireland. In 1926, 28 per cent of 16 and 17 year old boys in the Irish Free State were neither in gainful employment nor in school, more than three times the proportion in Scotland (8 per cent) and England and Wales (9 per cent), and substantially higher than Northern Ireland (17 per cent). Although the problem was most pronounced in rural areas, over 40 per cent of 16 and 17 year old boys, and over 60 per cent of 16 and 17 year old girls in the cities of Cork, Waterford and Limerick had no gainful employment; the corresponding figures for Belfast were 17 and 21 per cent, comparable to English and Scottish cities.[3] While emigration from Irish towns and cities was not unknown, it would appear to have formed a less common part of family survival strategy than in the countryside and provincial towns, with the result that poverty in the larger cities remained a persistent element in the social and socio-medical history of independent Ireland.

In the remainder of this article I describe and analyse three major phenomena: changes in life expectancy and mortality from the 1920s until approximately 1970; the comparative rates of urban and rural mortality and how these changed between 1920 and 1970; and the comparative rates of male and female mortality and sex differentials, as well as changes in the female advantage. The analysis will take account of comparisons between Ireland and other western countries, and the issues that the data raise with respect to the organisation of healthcare and social services. Where appropriate, comparisons will also be made with Northern Ireland. The article ends by setting out an agenda for future research into the social history of medicine in the early years of the independent Irish state, based on the issues highlighted here.

Irish mortality, c. 1920–1950

During the first half of the twentieth century, Ireland failed to keep pace with the improvements in life expectancy recorded in other developed countries. Nineteenth-century cities were extremely unhealthy places, so it is not surprising that life expectancy in Ireland, a predominantly rural country, exceeded life expectancy in the USA and England and Wales from 1864, when the first Irish registration data was collected, until the 1920s. By 1900, however, female life expectancy in the USA and in England and Wales had outstripped that in Ireland. US males recorded a higher life expectancy than Irish men in the 1920s; with England and Wales following suit in the 1930s.

The *Commission on Emigration* (1948–54)[4] published a statistical table giving crude death rates for 22 countries in Europe and the Caucasian world.[5] For the years 1926–35 Ireland had the fifth highest death rate of the 22 countries listed; for 1946–51 Ireland's was the highest. The *Commission* dismissed Ireland's poor showing by noting that 'Crude death-rates are not, however, a satisfactory measure of mortality because they do not take into account the age-composition of the population'. It claimed that comparative data on life expectancy presented the Irish figures in a much better light, with the country lying 13th of 22 countries for male life expectancy at birth.[6] However the supporting data was buried in an appendix, and data for Ireland in the years 1945–47 is compared with figures for Italy in 1931, Austria in 1933 and Northern Ireland between 1936 and 1938, in an apparent effort to present Ireland in the best possible light. When we examine only those countries where data related to the years after 1945, Ireland had the second lowest life expectancy of those cited.[7]

Irish life expectancy rose during the first half of the twentieth century (see Table 11.2), but the rate of improvement was modest. By 1950 Ireland had been overtaken in life expectancy by Northern Ireland, France and Switzerland. Life expectancy in Scandinavia and the Netherlands was already superior to Ireland by 1920. By 1950 Italy had a higher life expectancy for women, and was rapidly closing the gap with respect to men (see Tables 11.2 and 11.3). The improvement in life expectancy during this period was primarily achieved by reduced mortality from the major infectious diseases and by improvements in maternal and infant mortality.[8] In Ireland some modest, if at times

Table 11.2 Life expectancy at birth in Ireland

	Males	Females	Female advantage
1910–12	49.3	49.6	0.3
1925–27	53.6	54.1	0.7
1935–37	58.2	59.6	1.4
1940–42	59.0	61.0	2.0
1945–47	60.5	62.4	1.9
1950–52	64.5	67.1	2.6
1960–62	68.1	71.9	3.8
1965–67	68.6	72.9	4.3
1970–72	68.9	73.5	4.6

Table 11.3 Life expectancy at birth in various European countries

	1920s (c. 1920–22 unless indicated)			c. 1950		
	Male	Female	Female advantage	Male	Female	Female advantage
Belgium	56.0	59.8	3.8	62.0	67.3	5.3
Denmark	60.6	62.6	2.0	69.8	72.6	2.8
France	52.2	55.9	3.7	63.6	69.3	5.7
Italy	49.3	50.8	1.5	63.4	67.3	3.9
Netherlands	61.9	63.5	2.6	70.9	73.5	2.6
Switzerland	54.5	57.5	3.0	66.4	70.9	4.5
England & Wales	55.6	59.6	4.0	66.4	71.5	5.1
N Ireland	55.4	56.1	0.7	65.6	68.8	3.3
Scotland	53.1	56.4	3.3	64.4	68.7	4.3

1928–32

1925–27

uncertain, advances were achieved during the 1920s and the 1930s in mortality from infectious diseases, and therefore in total mortality, but these gains were reversed during the years of World War II, or 'the Emergency' as it was styled in neutral Ireland.

1. Tuberculosis

The belated improvement in mortality from tuberculosis was a major factor in the slow decline in overall mortality. As Greta Jones has shown, the TB epidemic peaked in Ireland later than in England and Wales, and the peak in the Irish Free State was later than in Northern Ireland. From the mid-1920s the death rate from TB in the Irish Free State was higher than in Northern Ireland. For the years 1927–37 the Irish Free State recorded the lowest annual percentage reduction in TB mortality of 20 countries in Europe, Australasia and North America'.[9] Nevertheless, mortality from TB was declining, albeit slowly until 1939, when the death rate rose in both Northern Ireland and the Irish Free State. By 1945 TB mortality in Northern Ireland had fallen below the 1939 figure, whereas in the Irish Free State it remained above that level until 1947.[10]

2. Other infectious diseases

The chronology for other infectious diseases, such as typhoid, typhus, scarlet fever and diphtheria, is similar. These diseases were identified as priorities for the local Medical Officers of Health, who were first appointed after 1927 (previously medical officers were only appointed in county boroughs).[11] Reducing the incidence of and mortality from infectious diseases entailed a combination of medical measures, such as immunisation against diphtheria and typhoid, and improvements in environmental health, including the provision of clean water and clean milk. Controlling typhus entailed delousing, as well as improvements in housing and sanitary services. The 1933–34 report of the Department of Local Government and Public Health noted that typhoid was an engineering problem as well as a medical one, and welcomed 'the triumph of sanitation' in urban areas (perhaps too confidently). The 1937–38 report claimed that typhoid was disappearing. Immunisation rates for diphtheria had shown a decided improvement; a pilot programme in Cork City Borough was regarded as particularly successful,[12] and there is some evidence that local authorities were making determined efforts to ensure that infants were immunised. John Brady, a Cavan farmer, recorded the vaccination of his first child, Mary Theresa, at the local dispensary in October 1932, when she was seven months old. In

Table 11.4   Major infectious diseases, 1940–44*

	1940	1941	1942	1943	1944
Notifications	4719	4173	6166	7349	7156
Deaths	385	273	268	381	425

Source: Report Department of Local Government and Public Health 1944–45)
*These diseases are typhoid, typhus, diphtheria, scarlet fever and puerperal sepsis.

August 1933 the family was contacted because of their failure to have their second child, Agnes Philomena (born 19 February 1933), vaccinated; she was then vaccinated on 4 October.[13] The incidence and mortality of the top five infectious diseases fell during the late 1930s. There were 273 deaths in 1941 compared with 687 in 1936. These trends were reversed in 1943 (see Table 11.4).

3. Maternal mortality

Puerperal sepsis was the other major infectious disease and is included in Table 11.4. From the late 1920s, the Department of Local Government and Public Health tried to combat it by promoting the medicalisation of childbirth with obstetric care, either in maternity hospitals or in the home by qualified doctors and midwives, and by discouraging the use of unlicensed midwives. By the 1930s the three Dublin maternity hospitals were providing ante-natal clinics, and the increasingly detailed annual reports published by the hospitals provided a basis for scrutinising obstetric practice and trends in mortality. Maternal mortality was also a major topic for research papers and lectures in Irish medical journals.[14] The proportion of Dublin births taking place in hospitals rose from 39 to 48 per cent between 1933 and 1938, and attendance at out-patient clinics almost doubled.[15] By the late 1930s a majority of Dublin mothers were being delivered either in maternity hospitals or in their homes by trained midwives and doctors attached to these hospitals. But despite these initiatives, mortality continued to fluctuate. The annual report of the Department of Local Government and Public Health for 1936–37 noted, rather despondently, that 'Notwithstanding ... the attention', there had been 'no substantial reduction in the death-rate'.[16] The timing of these remarks is ironic because 1936 saw the introduction of sulphonamide to Ireland, with almost immediate impact. By 1944, in a pattern that is completely contrary to trends for other infectious diseases, mortality from puerperal sepsis was one-third the rate for the years 1934–43.[17]

4. Infant mortality

In 1927–28 the Department of Local Government and Public Health described Irish infant mortality as 'moderate', while noting that infant mortality in urban areas was double the national average. By 1930–31 the department was less sanguine; it noted that Ireland compared unfavourably with England and Wales, but was decidedly better than Scotland. By the 1930s they had come to regard infant mortality as largely preventable. Attention was increasingly focused on the relationship between infant mortality and hygiene, maternal health and breast-feeding.[18] Dr James Deeny, who was appointed as Chief Medical Officer in the Department of Local Government and Public Health in 1944,[19] noted that mortality from enteritis in England and Wales fell from 1905, whereas in Ireland mortality peaked in 1915, fell from 1916 until 1925, but rose gradually from that time, peaking during the Emergency. Writing in 1946 he suggested that 'the rise in the enteritis death rate of infants during recent decades in Ireland took place at a time of great improvement in living conditions and that it was accompanied by a decrease in breast-feeding'.[20] The incidence of breast-feeding in Ireland and the timing of its decline remains a topic for further research, although there is substantial impressionistic evidence that suggests that bottle-feeding was already widespread by the 1930s.[21] Although infant mortality fell sharply from the late 1940s, the high incidence of gastro-enteritis among Dublin infants remained a cause for concern.[22]

One of the most disturbing aspects of Irish infant mortality is the extremely high death rate for illegitimate children; in 1930 one illegitimate child in four died in the first year of life, four times the mortality rate for legitimate children, and significantly higher than the mortality of illegitimate children in Northern Ireland (140 per 1,000) and England and Wales (105 per 1,000). These concerns led to the establishment of mother and baby homes, where pregnant single mothers awaited confinement. Most mothers stayed in the home for at least a year, caring for their child and carrying out unpaid domestic work. Most children remained in the home awaiting adoption; if not adopted they were boarded out until the age of seven when they were placed in an industrial school. Legislation providing for adoption was not introduced until 1952, though some private adoptions took place before that date.[23] These homes, which were run by religious orders were funded by the Irish Hospitals Sweepstake, a sweepstake held on the English classic horse races, which was established in the late 1920s to raise funds for Irish medical institutions.[24] Nevertheless, mortality of illegitimate children remained disturbingly high up to 1950: at 78 per 1,000 compared with 45 per 1,000 for children born in wedlock.[25] We cannot exclude the possibility that these figures may conceal cases of infanticide.

5. Urban vs rural mortality

During the nineteenth century the countryside was much healthier than cities and large towns because of the lower incidence of infectious diseases and illnesses associated with adverse environmental factors such as dirt, disease and contaminated food. In the twentieth century adverse urban mortality persisted longest in less-developed countries. In the USA infant mortality in urban areas fell below the rural rates in 1929 and kept declining until 1940; the urban penalty disappeared to be replaced by a penalty associated with class rather than geography.[26]

In Ireland the comparative advantage of rural over urban areas proved to be long-lasting. Regional life tables in 1941 showed that life expectancy was almost ten years greater for a baby boy born in Connacht (the most rural and the poorest province), than one born in Dublin; for girls the gap was seven years. The differential in percentage terms was greatest for those aged 45: a Connacht man could expect to live five years longer than his Dublin counterpart, a Connacht woman 2.75 years. Male mortality at the age of 55 was 23.4 per cent higher in Dublin than in Connacht; female mortality 13 per cent higher.[27] Age-specific mortality data by county and county borough for all Ireland (north and south) for the years 1963–67, calculated by Jacques Verrière, showed that mortality was highest in the most urbanised areas. In Northern Ireland, however, excess urban mortality only applied in the large cities, especially Londonderry, whereas in the Republic of Ireland excess urban mortality applied even in small towns.[28] Excess infant mortality offered a very imperfect explanation for excess urban mortality; Verrière concluded that the explanation should be sought in post-infant deaths. The greatest contrast was, as Barry stated in 1941, in mortality for those aged 40–65, and especially the age group 50–64.[29]

Accounting for trends in Irish mortality

1. 'The emergency'

Up to the outbreak of World War II government efforts to reduce mortality concentrated on infectious diseases, and on maternal, infant and child health. If we except immunisation against diphtheria and typhoid, and efforts to reduce puerperal infection by improving the obstetrical care, most advances in public health were dependent on improvements in diet, housing and the sanitary environment, and upon measures to isolate and to care for tuberculosis patients. Adverse mortality in urban areas was overwhelmingly a reflection of poverty and environmental circumstances.

The modest improvement achieved in mortality during the 1930s was reversed during the war years. Between 1941 and 1945, mortality rose in both rural and urban areas. In Britain civilian deaths showed 'no radical change' during the war years, though infant mortality rose during the early years, only to fall below pre-war levels by the end of the war, and the death rate among those aged over 65 was higher than in pre-war years.[30] Despite the war, and the disruption posed by mass mobilisation, the significant displacement of people, and the disruption from bombing raids, Britain appears to have been more successful in maintaining basic services and resources than neutral Ireland. Gerard Fee has documented the deterioration in health and living conditions in war-time Dublin. Unemployment rose because factories, building sites and the docks were badly hit by shortages of supplies. Fuel was scarce, as were many basic foodstuffs, especially bread, which the government decided not to ration. The switch from coal to turf as the most common fuel, and the sharp reduction in access to gas, brought about a deterioration in air quality in Irish cities. The high extraction rate of flour led to a significant increase in childhood rickets. In Britain millers were required to add calcium carbonate to flour to counteract this, but they were not in Ireland.[31] War-time shortages exacerbated long-standing inadequacies in the diet of the Dublin poor. Data from the National Nutrition Survey (1946–48), showed that in Dublin 'Among the poorer and larger families, a high proportion of the meals consisted simply of "bread and spread"'; the proportion of such meals declined in smaller and more prosperous families.[32] An associated clinical study revealed that 17 year old Dublin boys and girls who were members of large families had lower weights than their peers. Inadequate diets also appear to have impacted on the health and well-being of poor families in provincial towns. A sample of school-children drawn from various towns indicated 'a definite tendency for the heights and weights to decrease as the family size increases'. While noting that the sample was small and might be subject to sampling variations, the report nevertheless concluded that 'it seems very probable that there is, in town areas, a variation in height and weight with family size'.[33]

Rural Ireland fared much better during the Emergency. Compulsory war-time tillage and war-time turf work meant that jobs were more plentiful than before, and shortages of fuel and food could be eased by informal means. Food shortages persisted until the end of the 1940s;

indeed, bread was first rationed in 1947, but data collected from farming households indicated that they were adequately fed. However, clinical data relating to height and weight of rural children indicated that those aged 8–10 were in a poorer nutritional state than other ages. Families in the poorest areas along the western seaboard recorded vitamin A deficiencies and a high proportion of meals consisting of tea and bread (as in the Dublin tenements), but much greater consumption of milk and butter.[34]

It is relatively simple to compile evidence showing that the quality of life deteriorated for city-dwellers during the Emergency; it is less easy to do this for rural Ireland. Yet statistics indicate a greater rise in mortality in rural areas than in the cities during the Emergency. This might be due to better certification of deaths. Between 1941 and 1942 the number of births jumped from 56,780 to 66,117, and at least part of the increase was probably attributable to improvements in registration, so that the infants could qualify for ration cards. Rationing might also have led to the better enforcement of death registration. If, however, as statistics indicate, rural mortality rose to a greater degree than in the cities, food shortages were not the primary factor, so what other explanations come to mind? The 1930s saw significant investment in housing in both rural and urban Ireland, and improvements in water and sanitary services, though the latter were mainly confined to cities and towns.[35] During the Emergency, improvements in social infrastructure came to a halt because of a shortage of supplies. Additional demands placed on local authorities, including the responsibility for harvesting turf, organising civil defence, and formulating contingency plans in the event of an invasion, reduced the time and resources available for public health. The 1940 county management act abolished county boards of health, transferring their duties to county councils. The disruption of long-established administrative structures may have adversely affected the delivery of public health services. The 1940 act also saw the appointment of county managers, who assumed many functions formerly the preserve of elected councillors, and the early years of the new county management system brought internecine conflict between managers and councillors that was not conducive to efficient local services.[36] Rising mortality from diphtheria might suggest a fall-off in immunisation; yet the Dublin epidemic coincided with a sharp rise in the numbers being immunised.[37]

The deterioration in health during the war years was a cause of concern in the public service and among sections of the medical profession. The Statistical and Social Inquiry Society of Ireland presented many papers on public health issues during these years. The decision to appoint a Chief Medical Adviser in the Department of Local Government and Public Heath by open competition (instead of by internal promotion), and the establishment of a separate Department of Health can be seen as responses to this health crisis.[38] But much of the focus of 1940s research was on health problems in Dublin and other major cities; the rise in rural mortality appears to have passed largely unnoticed.

2. After the war

In both absolute and comparative terms Ireland fared better in the decades immediately after World War II, catching up on many countries that had overtaken it in the preceding decades in terms of life expectancy. Robert Kennedy remarked that 'The great improvement in life expectancy in Ireland between 1946 and 1961 was accompanied, ironically, by the highest rates of emigration out of Ireland since the 1880s'.[39] Verrière notes that in 1950 Irish infant mortality was high compared with other northern European states, but significant advances had been made by 1968, when infant mortality in Ireland was superior to that in Northern Ireland.[40] By the late 1960s infant mortality in Dublin was lower than in Belfast, whereas from 1900 to the mid-1960s infant mortality in Belfast had been consistently lower than in Dublin.[41]

3. Medical versus social intervention

The better performance after 1950 coincided with major advances in drug treatment and medical services, and a significant expansion in social spending. It also coincided with record levels of emigration, and with the lowest increase in GNP and material living standards in western Europe.[42] The incidence of puerperal sepsis fell sharply in the early 1940s – against the trend for overall mortality – because sulphonamides became widely available. The sharp fall in total mortality from the late 1940s coincides with the availability of penicillin and streptomycin to tackle rheumatic fever, pneumonia and TB. The years 1948–58 saw a sharp increase in public spending on health services, which contributed to a major programme of hospital construction, the appointment of specialists to regional hospitals, and the introduction of state medical services for mothers and infants.[43] Deeny and Barrington both attribute improvements in life expectancy during the 1950s to the rapid rise in spending on health, which should probably not surprise us since both have served as senior officials in the Department of Health.[44] However, the post-war housing programme, the expansion of social insurance and the introduction of children's allowances were of equal if not greater significance.[45]

Emigrants' remittances from Britain also became a significant source of income for many households, especially in the west of Ireland.

Despite the improvement in overall mortality, Irish life expectancy continued to lag behind that of other developed countries, and the urban disadvantage persisted, though at a lesser rate. The funding and organisation of health services and associated housing, sanitation and welfare services may be a factor. Dublin and Cork cities both faced major hurdles in financing housing and sanitary services, because they were required to raise their own funds without any assistance from the Exchequer, and there is evidence that this retarded major capital spending programmes, such as social housing schemes, especially during the 1950s.[46] Verrière suggests that the reason why urban disadvantage applied even in small towns in the Republic of Ireland, but not in Northern Ireland, was because Northern Ireland had a more developed social security system.[47] Another possible explanation may lie in the poor state of sanitary conditions in small towns in the Republic of Ireland. In the late 1950s most villages in county Limerick, for instance, were still without piped water and sewerage; the position would have been similar throughout the state.[48] Most farm houses also lacked ready access to running water and sanitary services, though the epidemiological consequences of this may not have been significant.[49]

Until the 1970s the primary funding for health services came from local authorities, and the inadequate and non-buoyant taxation base of Irish local authorities constrained the expansion of services and the development of new programmes. Irish farmers followed the example of UK farmers in clamouring for de-rating of agricultural land in the late 1920s; although they only achieved a partial concession, unlike UK farmers, the fact that Irish local taxes were wholly dependent on agricultural rates, and that farmers were one of the most effective lobby groups in the country and were not particularly concerned about health matters, must be considered.[50] Hence additional spending on health was dependent on outside resources. Hospitals, and mother and baby homes were funded by the Irish Hospitals Sweepstake, a source of funding that collapsed during the war, but then recovered, providing significant funds until the late 1950s.[51] Campaigns to transfer the cost of health services to central taxation began to emerge at that time. However, health spending ranked low in government priorities during the late 1950s and the 1960s. Irish membership of the European Economic Community was widely seen as offering the opportunity to raise spending on health and welfare, because farm price supports – the most expensive item in the budget by 1970 – would be borne by Brussels.[52]

4. Female advantage

One of the most interesting and under-researched aspects of Irish mortality is the extremely slight female advantage in terms of life expectancy. The graphs of comparative male and female life expectancy between 1920 and 1970 (see Tables 11.2 and 11.3) show that Irish women were disadvantaged relative to women in other European countries. According to Robert Kennedy, 'excess female over male life expectancy between 1817 and 1946 was less than Ireland than in England; females lived from three to five years longer than males in England but only 0.1 to 1.9 years longer in Ireland'.[53] Verrière emphasised that the low female advantage in Ireland reflected the relative disadvantage of women, not as in Sweden where men and women both had very favourable life expectancies.[54] He describes Ireland's poor female advantage as a vestige of the nineteenth century, and as comparable to patterns in Third World countries. This is correct because sex differences in mortality that favour women have become larger as mortality has declined.[55] Kennedy cites mortality data from the Registrar General for 1931 showing higher male mortality up to age nine in rural and urban areas, but excess female mortality from age ten to 54 in rural areas.[56] Female mortality in Ireland improved relative to men, between the 1940s and the 1950s, and this improvement persisted into the 1960s and later decades. Greta Jones notes that, up to the 1930s, women in Ireland had a higher rate of mortality from tuberculosis than men but that, after that time, male mortality was greater than female mortality.[57]

The relative disadvantage of Irish women in mortality terms has not to date attracted attention from researchers in women's history. This is not unusual, because it is only recently that sex differences in mortality and morbidity have been studied to any great extent.[58] Preston, one of the first scholars to examine this question in detail, notes that higher female mortality (relative to men) is associated with overall high mortality, and it declines as mortality levels improve. In societies with high female mortality, infant boys have a higher mortality than girls but, from ages one to 30, 'females in the group of populations with lowest life expectancy have higher average deaths from all causes combined' and, indeed, from practically every cause of death except violent deaths. By the age of 40 or so, life expectancy for women exceeds that of men in these societies.[59]

The relative disadvantage of Irish women was very much a rural phenomenon: in urban Ireland the difference in life expectancy between men and women was similar to that in England and Wales. Preston maps differences in mortality by gender over a wide range of countries, and

notes that the three countries whose pattern differs from 'culturally affiliated neighbouring countries', namely, Ireland, New Zealand
and Chile, all have significantly different levels of urbanisation than
their neighbours. This suggests that in areas where other factors are
broadly similar, higher than expected female mortality can be found in
predominantly rural societies.[60]

Explanations for female advantage or disadvantage break down into
biological and environmental factors. It has been suggested that women
suffer higher mortality from infectious and parasitic diseases than men,
and that the emergence of female advantage in life expectancy reflects
the conquest of these killers,[61] but as infectious diseases were much more
prevalent in Irish cities and towns than in the countryside, this does not
get us very far in explaining the Irish statistics. Indeed, it suggests that
the rural/urban gender disadvantage might have been even greater than
the statistics suggest. Mention high female mortality and someone will
invariably raise the issue of Ireland's high marital fertility, but although
maternal mortality in Ireland remained above the level in England and
Wales until the late 1960s, the number of deaths was not of a scale to
have had a significant impact on comparative life expectancy data: maternal mortality in 1938 was 234 (the total number of deaths among
women was approximately 20,000). Moreover, in the 1920s and 1930s
the relative disadvantage of women in rural Ireland, compared with
women elsewhere or with women in urban Ireland, began at approximately ten years of age, which is consistent with the pattern in the Third
World today.[62] Kennedy cites statistics for deaths of boys and girls aged
5–14 years in England and Wales, USA and Ireland, revealing that in
Ireland female death rates from TB were 640 per cent of the male rate,
whereas in the USA and England and Wales male and female death rates
from TB were identical; for rheumatic fever, rheumatic heart disease and
scarlet fever, the female death rates in Ireland were 270–290 per cent of
the male; whereas in the USA and England and Wales, death rates were
generally identical , except in the case of rheumatic heart disease, where
female mortality was double that of the male.[63]

Verrière ventures to suggest that if high fertility was a contributory
factor in the poor female advantage of Irish women, it had a delayed
impact, manifested in diseases and illness in later years,[64] but this argument is not supported by Preston's work, or by Irish data showing that
for older age groups death rates in rural Ireland were lower for women
than for men. A cross-country analysis of differences in female advantage, carried out over 169 countries, showed that female advantage
rose by 0.44 years for each unit decrease in the number of children, but

the author noted that fertility was highly correlated with female illiteracy, and wondered whether illiteracy rather than fertility might be the explanatory factor.[65] Female illiteracy in Ireland was comparable if not lower than male illiteracy throughout the period under study, and women had equal and at times better access to elementary schooling, so this does not advance our understanding.

In Ireland provisions for maternal healthcare, or their absence, may have had a long-term impact on adult female mortality. Women in urban areas were more likely to receive ante-natal care and obstetric care by medical professionals, and this may have had long-term benefits in terms of health, since such care commonly involved an assessment of overall health, and would reveal serious medical problems, such as diabetes, cardiac and pulmonary diseases, possibly leading to their successful treatment. There was no significant urban/rural difference in terms of fertility; farmers and unskilled labourers had the largest families.[66]

The widely accepted position in international literature is that differences in female life advantage relate to social and cultural factors.[67] In parts of the world with strong Hindu or Confucian traditions, preference for sons is undoubtedly a major factor, since this results in the earlier weaning of female babies, and the provision of less food, less healthcare and less education for girls.[68] Nowadays this is reflected in differential birth rates, with females much more likely to be aborted than males in India or China. Female advantage internationally shows a negative correlation with the proportion of the population in agriculture.[69] Kennedy notes that the years of greatest excess female mortality in Ireland were the time of changing land laws; he suggests a possible link with land succession and property transfers,[70] implying that farmers privileged their sons and may have disadvantaged their daughters in order to secure a male heir, and may have been less concerned about the survival of female children, who would have had to be dowered, which was considered to be a drain on family resources. He also concludes that 'males were dominant in Irish society, and that they controlled many of the resources needed for good health in rural areas'.[71] Among the evidence of male dominance that Kennedy cites is Arensberg and Kimball's description of women waiting on men while they ate, and then eating later. Can we be certain that there was always sufficient food left for them, or that the men had not eaten all the bacon, eggs or whatever protein was available? There is substantial evidence that in Dublin tenement households c.1909, men got most of the protein available, and children got most of what was left, leaving women to survive on bread and tea.[72] Did a similar, or even worse, division of food apply in rural Ireland? We have no evidence, since the National Nutrition Survey collected data on meals per household; it made no effort to look at how the food was distributed.

One indirect source of evidence regarding the inferior treatment of female children in rural Ireland comes from statistics on admissions to industrial schools. Most children committed to industrial schools were committed because their families (nuclear or extended) were no longer able or willing to care for them, or were perceived as unable to care for them. Girls always constituted a majority of those committed because of poverty and related factors; they were committed at an earlier age than boys, and they stayed longer in these institutions. This suggests that when a family came under pressure, boys were more likely to be kept within the family than girls. However, the gender ratio for Dublin children committed to industrial schools was significantly different to the national picture; in July 1950, for example, 988 boys and 817 girls from Dublin City were inmates of industrial schools and reformatories, compared with 1,831 boys and 2,348 girls with addresses outside Dublin.[73]

Economic factors cannot be excluded from the analysis of female health and mortality. The heavy manual work carried out by women in farming and associated jobs has been suggested as one factor in the relatively worse mortality of rural women, but Irish women did not carry out much heavy outdoor work except in the remote areas along the western seaboard. Indeed the dearth of work for women on Irish farms may have been a more critical factor. There was very little paid work for women in rural Ireland, and many rural women may have been regarded as surplus to requirements. Every farm needed one woman to cook, clean and mind the hens but one was probably sufficient, whereas there was often work for two or more men. Rural men could supplement their income by claiming unemployment assistance, whereas women were generally denied this form of income supplement, with the exception of widows, who were better treated than widowers in terms of state support. By the mid-twentieth century, there were many diatribes about 'the flight of the girls' from rural Ireland; available information on comparative mortality suggests that migration from rural Ireland to cities and towns in Ireland or overseas would have been a rational decision for Irish women. In 1841, on the eve of the famine, women accounted for precisely 50 per cent of the rural population but, by 1901, there were only 961 women for every 1,000 men and, by 1951, the ratio had fallen to 868, compared with 1,120 : 1,000 in urban areas. The gender imbalance was even more pronounced among younger adults with 1,165 women per thousand men in the age group 20–29. But the diatribes against the lack of women in rural Ireland rarely took account of the fact that, whereas life expectancy for men was greater in the countryside than in the cities, the reverse applied for women. Again these statistics were not hidden; they were readily available, but they were not seen as a matter for concern.

At this stage the best that we can do is flag this as a topic that urgently requires further research. Some possible areas for investigation include milk-borne diseases, such as TB and typhoid, because women were traditionally responsible for milking cows, and bovine tuberculosis remained rife in Irish herds until the 1960s. We also need to identify factors that might make women more susceptible to acute respiratory diseases and rheumatic fever. There is evidence that many Irish women were sent back from the US in the 1920s because they suffered from heart disease caused by rheumatic fever.[74] However the explanation may lie deeper in Irish society. With an average of six children per farming family, surplus daughters may have rated very low in the allocation of food, medical care and general nurturing. Alternatively, we should conclude that life in the countryside was good for men.

Future research

An understanding of the underlying factors in the three issues that I have outlined will require a more detailed and much more rigorous analysis of the statistical information, from both Irish and comparative sources. With regard to female advantage (or lack thereof), a detailed analysis of mortality by gender might prove illuminating. But the answers to these questions, even partial answers, will not be found in medical statistics or in the history of medical institutions, but through the integration of medical knowledge with our knowledge of the state and society. Some of the issues that arise include deciphering the priorities that directed government spending, the incidence of taxation (whether spending was a national or a local charge), the degree to which the medical profession might influence the investment of government spending on health towards hospitals, doctors and nurses rather than other forms of spending, such as preventative medicine, and of course the attitudes of the state, church and society towards compulsion and state regulation, which was so much at the core of the Mother and Child debate.[75] Other issues involve the wider ideological attitude towards Irish cities and city-dwellers, which meant that high urban mortality was not seen as a serious cause for concern, rather than as an argument for measures to sustain rural Ireland. Similarly an understanding of the low level of female advantage in rural Ireland will entail a close analysis and understanding of the power relations within the rural family and rural society. A final and tantalising topic for research relates to the long-term impact of famine and food shortages on the Irish population. Research by Cecily Kelleher and her team on Irish immigrants in the USA indicates an increased risk of cardiovascular disease, compared with Italian immigrants; Kelleher surmises that this may reflect the impact of surviving the famine on the Irish gene pool.[76]

Notes

1   Unless otherwise stated, in this paper when I say Ireland I mean the 26 county state that became the Irish Free State in 1922.

2   In 1911 General Labour accounted for 17 per cent of occupied males in Belfast and 19.5 per cent in Dublin, compared with 2.9 per cent in Glasgow and 2.5 per cent in Edinburgh. Mary E. Daly, 'Working class housing in Scottish and Irish cities on the eve of World War I', in S.J. Connolly, R.A. Houston and R.J. Morris (eds), *Conflict, Identity and Economic Development: Ireland and Scotland, 1600–1939* (Carlisle, 1995), p. 218.

3   *1926 Population Census*, vol. 9, pp. 143–6, 97–100.

4   *Commission on Emigration and other Population Problems*, 1948–54. Irish Government Official Publications, P. 25411 R. 63, Majority report. The Commission was set up to investigate past, current and future trends in Ireland's population and to make recommendations on measures that might be taken to influence these trends. The Commission's primary concerns were emigration and Ireland's low rate of marriage and late age of marriage. Chapter 6 'Deaths' accounted for 12 of the 190 pages in the report.

5   Comparative data was given for England and Wales, Scotland, Northern Ireland, Austria, Belgium, Denmark, Finland, France, Italy Norway, Portugal, Spain, Sweden, Switzerland, USA, Australia, New Zealand (excluding Maoris), Canada, and Union of South Africa (Europeans).

6   *Commission on Emigration*. Majority report, paras 227 and 231.

7   *Commission on Emigration*, Statistical Appendix, Table 23. Data on life expectancy post-1945 was given for Australia, Belgium, Canada, Denmark, England and Wales, Netherlands, Norway, Scotland and the USA. Only the USA recorded a lower life expectancy at birth for males than Ireland.

8   Henry F. Dowling, *Fighting Infection: Conquests of the Twentieth Century* (Harvard, 1977).

9   Greta Jones, *'Captain of all these Men of Death': The History of Tuberculosis in Nineteenth and Twentieth-Century Ireland* (New York and Amsterdam, 2001), pp. 135–6.

10 Ibid., p. 187.

11 The first county Medical Officer of Health was appointed in Kildare in January 1928; the last county to appoint an MOH was Laois in 1935. Michael P. Flynn, *Medical Doctor of Many Parts* (Dublin, 2002), p. 15.

12 *Annual Report, Department of Local Government and Public Health (DLGPH)*, 1933–34, pp. 47, 57; *Annual Report, DLGPH, 1937–38*, p. 41. *Annual Report, DLGPH, 1932–33*, p. 44 (re Cork diphtheria).

13 John Brady Diaries, 1921–75 in author's possession.

14 Bethel Solomons, 'Methods of obstetrics diagnosis at the Rotunda in 1909 compared with 1929', *Irish Journal of Medical Science*, Sixth Series, 73 (February 1932), pp. 67–75; G.J. Tierney, 'Maternal mortality', *Irish Journal of Medical Science*, Sixth Series, 82 (October 1932), pp. 603–6; Bethel Solomons, 'The prevention of maternal morbidity and mortality', *Irish Journal of Medical Science*, Sixth Series, 88 (April 1933), pp. 171–8; A.H. Davidson, 'The future treatment of puerperal sepsis in Dublin', *Irish Journal of Medical Science*, Sixth Series, 150 (June 1938), pp. 257–69.

15 Ruth Barrington, *Health, Medicine and Politics in Ireland, 1900–1970* (Dublin, 1987), p. 131.

16 *Annual Report, DLGPH, 1936–37*, p. 71.

17 *Annual Report, DLGPH, 1944–45*, p. 31; Lindsey Earner-Byrne, *Mother and Child. Maternity and Child Welfare in Dublin, 1922–1960* (Manchester, 2007), pp. 37–9.

18 *Annual Report, DLGPH, 1927–28*, p. 41; *Annual Report, DLGPH, 1930–31*, pp. 63–4; *Annual Report, DLGPH, 1931–32*, p. 56.

19 For the background to Deeny's appointment see Barrington, *Health, Medicine and Politics in Ireland*, pp. 155–6.

20 James Deeny, *The End of an Epidemic: Essays in Irish Public Health 1935–65*, ed. by Tony Farmar (Dublin, 1995), pp. 60–1. Extracted from 'The epidemiology of enteritis in Ireland', *Journal of the Medical Association of Eire* (October 1946).

21 When the Department of Agriculture proposed establishing a factory to manufacture powdered milk in 1937, the Minister remarked on the need to reassure Irish mothers about the quality of the milk formula, and for that reason the well-known British brand, Cow and Gate, was manufactured under licence. See Mary E. Daly, *The First Department: A History of the Department of Agriculture* (Dublin, 2002), p. 176. All the children of Cavan farmer John Brady, (born from 1932) were bottle-fed, as indicated by accounts showing the purchase of baby bottles, teats and infant formula. On breast-feeding see also Catriona Clear, *Women of the House: Women's Household Work in Ireland, 1922–1961* (Dublin, 2000), pp. 126–42 and Earner-Byrne, *Mother and Child*, pp. 63–5.

22 J.H. Elwood, 'Infant mortality in Belfast and Dublin 1900–1969', *Irish Journal of Medical Science*, 142 (1973), p. 172; Medical Research Council of Ireland, *Report of the Gastro-enteritis Survey Dublin City 1964–66* (Dublin, 1971).

23 Mary E. Daly, *The Slow Failure: Population Decline and Independent Ireland, 1920–1973* (Madison, 2006), pp. 286–7; Earner Byrne, *Mother and Child*, pp. 172–92.

24 Mary E. Daly, '"An atmosphere of sturdy independence": The state and the Dublin hospitals in the 1930s', in Greta Jones and Elizabeth Malcolm (eds), *Medicine, Disease and the State in Ireland, 1650–1940* (Cork, 1999), pp. 238–49.

25 *Commission on Emigration*, paras 241–2.

26 Herbert S. Klein, *A Population History of the United States* (Cambridge, 2004), p. 152.

27 Colm A. Barry, 'Irish regional life tables', *Journal of the Statistical and Social Inquiry Society of Ireland*, 16 (1941–2), pp. 1–18.

28 Jacques Verrière, *La population de l'Irlande* (Paris, 1979), p. 349.

29 Ibid., p. 353.
30 Christopher Lawrence, 'Continuity in crisis: Medicine, 1914–1945', in W.F. Bynum, Anne Hardy, Stephen Jacyna, Christopher Lawrence and E.M. Tansey (eds), *The Western Medical Tradition 1800 to 2000* (Cambridge 2006), p. 388.
31 On rickets and its association with phytic acid and high extraction rates for flour, see Gerard Fee, 'The effects of World War II on Dublin's low-income families' (Ph.D thesis: University College Dublin, 1996), pp. 170–5.
32 Department of Health, *National Nutrition Survey, Part I. Dietary Survey* – Dublin Sample, K 53 (1948), p. 16.
33 Department of Health, *National Nutrition Survey, Part VII. Clinical Survey*, pp. 11–13.
34 *National Nutrition Survey, Clinical Survey, Part VII*, K 53–7; *National Nutrition Survey II, Congested Districts*, K 53–2.
35 Mary E. Daly, *'The Buffer State': The Historical Roots of the Department of the Environment* (Dublin, 1997), pp. 154–248. The Poulaphouca water scheme brought significant improvements in water supplies throughout the Dublin area.
36 Daly, *The Buffer State*, pp. 300–5, 311–15.
37 *Irish Journal of Medical Science* (1942), Typhoid in Cork city, p. 157 (1943); Diphtheria in Dublin, pp. 97, 140 (1944); Diarrhoea and enteritis in Dublin, pp. 25 and 111.
38 On these topics see Mary E. Daly, *'The Spirit of Earnest Inquiry': The Statistical and Social Inquiry Society of Ireland, 1847–1947* (Dublin, 1997), pp. 120–2; Barrington, *Health, Medicine and Politics in Ireland*, pp. 137–66; James Deeny, *To Cure and to Care: Memoirs of a Chief Medical Officer* (Dublin, 1989).
39 Robert E. Kennedy, Jr., *The Irish: Emigration, Marriage, and Fertility* (Berkeley, 1973), p. 48.
40 Verrière, *La population de l'Irlande*, p. 347.
41 Elmwood, 'Infant mortality Belfast and Dublin', pp. 166–8.
42 Kennedy, *The Irish*, p. 48.
43 Although the original Mother and Child Scheme providing for free medical care for expectant mothers and their children to the age of 16 collapsed because of opposition from the Catholic church and the medical profession, a modified scheme providing up to six ante-natal visits for expectant mothers and six weeks' post-natal care for mothers and infants was introduced in 1953. See Earner-Byrne, *Mother and Child*, pp. 134–44.
44 Barrington, *Health, Medicine and Politics in Ireland 1900–1970*, pp. 248–9; Deeny, 'How the Irish died in 1954', in *The End of the Epidemic*, ed. by Farmar, pp. 137–8.
45 For details see Daly, *The Buffer State*, pp. 321–79. The importance of redistributive social programmes is highlighted by Simon Szreter, 'The McKeown Thesis. Rethinking McKeown: The relationship between public health and social change', *American Journal of Public Health*, 92:5 (May 2002), pp. 722–5. In 1944 children's allowances were introduced, though only for the third and subsequent children; this was extended to the second child in 1952. There was a major extension of social insurance provisions under the 1952 Social Welfare Act. See Desmond Farley, *Social Insurance and Social Assistance in Ireland* (Dublin, 1964), pp. 56–7, 71–4.

46 Daly, *The Buffer State*, pp. 195, 246–56.

47 Verrière, *La population de l'Irlande*, pp. 349–53.

48 Jeremiah Newman (ed.), *Limerick Rural Survey*, Appendix B, Index of Social Provision, p. 310. Limerick had a poor record for immunisation against diphtheria; between 1951–55 only 70 per cent of children were immunised, whereas other counties achieved 100 per cent immunisation. Diarmaid Ferriter, *Cuimhnigh ar Luimneach: A History of Limerick County Council 1898–1998* (Dublin, 1998), p. 147.

49 Mary E. Daly, '"Turn on the tap": The state, Irish women and running water', in Maryann Valiulis and Mary O'Dowd (eds), *Women and Irish History* (Dublin, 1997), pp. 206–19.

50 On local taxation see Daly, *The Buffer State*, pp. 500–13.

51 Daly, 'An atmosphere of sturdy independence', p. 249.

52 These sections draw on Daly, *The Buffer State* and idem., *The First Department*.

53 Kennedy, *The Irish*, p. 55.

54 Verriere, *La population de l'Irlande*, p. 354.

55 C.A. Nathanson, 'Sex differences in mortality', *Annual Review of Sociology*, 10 (1984), pp. 191–2. In 1975–78 median difference male and female life expectancy at birth was 6.4 years in developed countries; in 1900 the difference was two years or less.

56 Kennedy, *The Irish*, p. 59.

57 Jones, '*Captain of all these Men of Death*', p. 67.

58 Nathanson, 'Sex differences in mortality', pp. 191–2 notes that the 1979 *Handbook of Medical Sociology* contains only a single reference to sex differences in mortality. Robert William Fogel, *The Escape from Hunger and Premature Death, 1700–2100: Europe, America and the Third World* (Cambridge, 2004), is a wide-ranging study that examines broad statistical trends but does not refer to female advantage or gender-based differences in mortality and morbidity.

59 Samuel H. Preston, *Mortality Patterns in National Populations with Special Reference to Recorded Causes of Death* (New York, 1976), pp. 121–2.

60 Ibid., pp. 138–41.

61 F.C. Madigan and R.B. Vance, 'Differential sex mortality: A research design', *Social Forces*, 35 (1957), pp. 193–9.

62 Nathanson, 'Sex differences in mortality', p. 195.

63 Kennedy, *The Irish*, pp. 62–3.

64 Verrière, *La population de l'Irlande*, p. 357.

65 Jean Lemaire, 'Why do females live longer than males?', *North American Actuarial Journal*, 6:4 (October 2002), p. 22.

66 Daly, *The Slow Failure*.

67 Jean Lemaire, in a cross-sectional regression study using data from 169 countries, showed that four population variables unrelated to biological sex differences explain over 61 per cent of the variability of life expectancy: number of persons per physician (a proxy for economic development); fertility rate; percentage of Hindus and Buddhists; and status of countries as former members of the Soviet Union. The current discussion of this issue is complicated by life-style factors such as smoking and drinking. Lemaire, 'Why do females live longer than males'.

68 Lemaire, 'Why do females live longer than males', pp. 11–12.

69   Preston, *Mortality Patterns in National Populations*.

70   Kennedy, *The Irish*, p. 61.

71   Ibid., pp. 54–5.

72   *Royal Commission on the Poor Law. Report on Ireland*, app. vol. x. 'T.J. Stafford, 'Notes on the social conditions of certain working class families in Dublin' (1910 ed.) 5070, l, 350.

73   *Report of the Department of Education* (Dublin, 1950), 1Pr9-50 Pr571, Table O, p. 183.

74   Francis P. Cavanaugh, 'Immigration restriction at work today. A study of the administration of immigration restriction by the United States' (Ph.D. thesis: Catholic University of America, Washington, D.C. 1928).

75   The Mother and Child Scheme was a government programme to provide universal free maternity and childcare as part of a reformed health service introduced in 1947. From the outset the proposal aroused opposition from the medical profession and the Catholic Church, and when Noel Browne, Minister for Health in the coalition government elected in 1948, attempted to implement the proposal he was not supported by his Cabinet colleagues. A modified version of the proposal providing for less comprehensive care and nominal charges for wealthier families was implemented in 1953. For details see John Whyte, *Church and State in Modern Ireland, 1923–79* (Dublin, 1980).

76   C. Cecily Kelleher, John Lynch, Sam Harper, Joseph B. Tay and Geraldine Nolan, 'Hurling alone? How social capital failed to save the Irish from cardiovascular disease in the United States', *American Journal of Public Health*, 94:12 (December 2004), pp. 2162–9.

Index

abortion, 210
Act to Regulate the Qualifications of
    Persons Holding the Office of
    Coroner in Ireland (1822), 48
alcohol abuse, 102
Amnesty Act (1885), 123
An Bord Altranais (Irish Nursing
    Board), 5
Anthropological Society of Paris, 120
Antrim district asylum, 103
apothecaries 19, 20, 22, 30
Apothecaries Hall, 20
Arnot, Margaret, 99
Ashe, Dr Isaac, Dundrum Asylum, 104
asylum, private, 86
    system, 86
asylums, admission to lunatic, 86–7
attempted suicide, 87

bacteriology, 156, 157
Bailie, Dr, Belfast, 197, 198
Balfour, Arthur, 51
Barr, Dr James, 50–1
Barrington, Jonah, 21
Barrington, Ruth, 219–20, 239
Barry, Edward, physician, 23
Barry, Jonathan, 148
Battersby, John George, M.D., 63
Belfast, 87, 98
    influenza in, 145
Belfast public health committee, 197,
    198
Belfast union hospital, 194
Belmullet, county Mayo, 71
Berkeley, Bishop, 28
Birrell, Augustine, 124
birth control
    and censorship, 210
    and Irish government, 210
    prohibition in Ireland, 207, 217
bleeding, 25
blistering, 26
Board of guardians, 194

Board of guardians, Longford, 154
bonesetters, 70
Bourke, Mr F., 107
Boyle union, 152
British Medical Journal, 43, 120
British Ministry of Health, 149
British Social Hygiene Council, 198
Broadmoor, 95
Broca, Paul, 120
Browne, Noel, 220
Byrne, Ellen (infanticidal mother),
    100

Cadden, nurse Mary, 187, 212
Cameron, Sir Charles, 155, 157
Campbell, Sarah, 85, 88
Carlow, 117
Carlow Lunatic Asylum, 71
Carrigan committee, 216
Cassell, Ronald, 61, 70
Casti Connubii, 211, 213, 215
Castlebar, 158
Castlecomer, Lord, 86
Cathcart, David, 85
Catholic Action, 208
Catholic church, 94
    alliance with medical profession,
        212
    and dispensaries, 71–2
    and flu, 158
    and medical profession, 207
    and suicide, 81
    position on contraception, 211
    reaction to mother and child
        scheme, 220
Catholic medical ethics, 211, 213
Catholic nurses, 209
Catholic Truth Society, 208, 216
Catholic University School of
    Medicine, 113, 116, 121, 123
censorship
    and birth control, 210
    of publications, 218

Censorship of Publications Act
    (1929), 210
Central Criminal Lunatic Asylum, 92,
    95, 108
Central Mental Hospital, 95
Charot, Jean Martin, 113, 119, 121,
    127
chemists, 70
Cheyne, George, physician, 29
childbirth, 183
Churchill, Dr Fleetwood, 101
Claremorris, county Mayo, 152
Clinical Society of Paris, 120
Clonakilty board of guardians, 153
Cloyne, Cork, 155
Coffey, Denis, 121
Colum, Mary, 124
Colum, Padraic, 124
Commission on emigration, 231
condoms, 219
Congested Districts Board, 59
Connolly, William, 16, 25
Contagious Diseases Acts, 200
contraception, 209
    and churches, 211
contraceptives
    banning of, 217
    sale of, 210
convict records, 94
Cork
    influenza in, 144
    number of medical practitioners
        in, 65
Cork guardians, 159
Cornwall, Elizabeth (infanticidal
    mother), 99
coroners, 37, 84
    appointment of, 41
    complaints regarding expenses, 46
    debates over practice, 42–4
    financial irregularities, 41
    fees, 45
    inquest, 81, 85
    numbers in Ireland, 41, 48
    role of, 39–40
Coroners' acts (1881, 1886), 41, 45,
    47
Corporation of Apothecaries, 20
Corrigan, Dominic, 121, 122

County Management Act
    (1940), 238
Crane, Robert J., LRCSI, 64
Cranfield, Dr Thomas, 63
Criminal Law Amendment Act
    (1935), 217
Criminal Law Amendment Act
    (1885), 216
criminal lunatic, 103
Crossman, Virginia, 3
Cullen, Paul, apothecary and
    accoucher, 71
Cullinane, Henry, 83
Cumann na mBan, 152

Damaged Lives, film, 198, 199,
    201–2
    church reaction to, 199
    local councillors reaction to, 199
Dangerous Lunatics Act (1838:
    1867), 86
Darwin, Charles, 114, 117, 121
de Boulonge, Duchenne, 119
de MacMahon, General Patrice, 118
de Valera, Eamon, 216
Deane, Seamus, 129
Dease, William, surgeon, 39, 40, 45
death sentence, for infanticide, 94,
    102, 178
Deeny, Dr James, 235, 239
Delaney, S.J., Edward, 117–18
Delgany dispensary district, 61, 67
Delagny management committee, 69
Department of Health, 197
Department of Local Government
    and Public Health, 194, 218, 233,
    234, 235, 239
Descent of Man, 121
diet, 158
Digby, Anne, 65, 68
diphtheria, 233
dispensary depots, 71
    access to, 67
dispensary districts, consolidation
    of, 66
    *General rules for the government of
        dispensary districts*, 63
dispensary management committees,
    67

dispensary doctors, 58, 86, 213
  and attendance, 66
  and Catholic clerics, 71–2
  and influenza, 144
  and medical authority, 73
  and pluralism, 66
  and residency, 66
  appointment of, 62
  distance from patients, 66, 68, 69–70
  reminiscences of, 69
  salary, 69
  visiting patients, 68
dispensary relief, 66
dispensary system
  and flu, 144, 152
  and management of, 60–1
  and poor, 61
  in Donegal, 62
  in Mayo, 62
  origin of, 60
  records of, 59
  relief tickets for, 61
dispensaries
  numbers of, 60
  numbers using, 61
domestic servants, 186
Donegal, number of medical
    practitioners in, 65
Doolin, William, 116
  reminiscence of George Sigerson, 124
Drennan, William, physician, 23
Dublin hospitals, 144
Dublin
  births in, 234
  coroners in, 41–2
  number of medical practitioners in, 65
Dublin Journal of Medical Science, 119
Dublin Medical Press, 37, 38, 46
Dublin Medical Press and Circular, 42
*Dublin Quarterly Journal of Medical
    Science*, 101
Duffy, Charles Gavan, 123
Dun, Lady, 28
Dun, Sir Patrick, physician, 22
Dundrum asylum, 92, 94, 95, 99, 100,
    101, 184
  discharge of patients, 97, 98
  numbers confined for infanticide,
    95, 96, 97

Dungarvan workhouse, 63
Dungloe dispensary district, 66
Durkheim, Emile, 82

Eagleton, Terry, 128–9
Egan, Dr Richard, Dublin city
    coroner, 42
'Emergency, The', 236
emigration, 64, 230
Enniscorthy
  and flu, 156
  dispensary, 71
Enniscorthy dispensary district, 67
eugenics, 209
executions, 94

Farrell, James, 86
fees to medical witnesses, 39
fenians, 114, 122
Fermanagh county council, 192
fever, 24, 26
ffrench, Hon. M J, magistrate, 84
Finnane, Mark, 86
Fissell, Mary, 14
Fitzgerald, Thomas Judkin, 83–4
flu, 159
  and alcohol, 159
  and class, 157
  and diet, 158
  and dispensary system, 144
  and geographical range, 150
  and home care, 152
  and individual responsibility,
    158
  and local government board, 151
  and medical advertisements,
    148–9
  and medical advice, 155
  and medical understanding,
    144–5
  and mortality, 142, 150
  and spread of, 155
  colloquial names for, 141
  epidemics, 142
  folk remedies, 159
  vaccines, 156
folklore, 105–6, 145
folk medicine, 14, 159
forensic psychiatry, 99

Freud, Sigmund, 113, 120
Furlong, Dr Nicholas, 63

Gaelic Athletic Association, 113
Gaelic medical schools, 14
Galway
    influenza in, 145
Gastro-enteritis, 235
Gardai, 184
General rules for the government of
    dispensary districts, 63
Geoghan, Bridget, 86
Gibbons, Luke, 128–9
Gilmartin, Archbishop of Tuam, 212
Glass, Mary (infanticidal mother),
    103
Glenties poor law union, 66
Gonne, Maud, 121
Government of Ireland Act
    (1920), 191
Grand jury, 46
Grimshaw, Thomas, 43
Guinness, Arthur, MD, 46

Hall, Marshall, neurologist, 121
handywomen, 70, 71
Head, Sir Henry, 132
health services, funding of, 240
herbalists, 70
history of medicine, 13
History of Irish Land Tenure and
    Land Classes (1871), 123
humoral medicine, 24
Hyde, Douglas, 115, 125, 126

illegitimacy, 93, 183, 187
    rates of in Ireland, 234–5
illegitimate children and mortality,
    235
industrial schools, children
    committed to, 244
infanticide
    and courts, 94
    and death sentence, 102, 178, 183
    and domestic servants, 186
    and insanity, 104, 184
    and law, 107, 177
    and mental health, 179, 185
    and medical opinion, 182

decline of, 94
definitions of, 92
in England, 95, 92
numbers of, 94
prison sentences, 106–7
profiles of women who
    committed, 96
reporting of, 94
sources of information, 108
Infanticide Act (1949), 92, 107,
    186
Infanticide Bill, 178
infant mortality, 235
infectious diseases, 234
influenza, 85, 159
    and alcohol, 159
    and class, 143
    and diet, 158
    and dispensary system, 144
    and geographical range, 150
    and home care, 152
    and individual responsibility, 158
    and local government board, 151
    and medical advertisements,
        148–9
    and medical advice, 155
    and medical understanding, 144–5
    and mortality, 142, 150
    and spread of, 155
    colloquial names for, 141
    epidemics, 142
    folk remedies, 159
    vaccines, 156
inquests, 81, 85
    format of, 39–40
    government interference, 49
    jury, 39–40
    of John Mandeville and Dr James
        Ridley, 49–52
insanity, 97
    and Catholic church, 81
    and law, 80
    and Protestant churches, 81
    and suicide, 80
    and women, 104
    temporary, 51, 79, 81
    verdicts of, 79
insanity defence, 106, 108
inspectors of lunacy, 97, 99, 103

Ireland
    coroners in, 41, 48
    demography, 229–30
    dispensaries in, 60
    doctor patient ratio, 65
    gender imbalance, 244–5
    institutions in, 57
    life expectancy, 231
    Lord Lieutenant in, 98
    mortality, 142, 150, 230–6
    nutritional standards, 237–8
    population decline, 95
    prison system, 95
    rates of suicide in, 82
    rural, 57
    Russian flu in, 143
    sanitation, 240
Irish Booklover, 124, 125
Irish Grand Jury Act (1836), 46
Irish Guild of Catholic Nurses, 208
Irish Guild of St Luke, SS Damian
    and Cosmos, 208, 213, 214, 215,
    220, 221
Irish Nursing Board (An Bord
    Altranais), 5
Irish Hospitals Sweepstake, 235, 240
Irish literary revival, 114
Irish Medical Association, 42, 43, 47
    dispute over medical witnesses,
    42–4
Irish Medical Directory, 42, 46, 48
Irish Quarterly Review, 48
Irish nursing board, 5
Irish prisons board, 51

Jacob, Archibald, chair of the
    Council of the Irish Medical
    Association, 42
Jackson, Mark, 99–100
Jones, Greta, 2, 3, 4, 5, 65
Journal of Mental Science, 82, 101, 103
Joyce, James, 122

Keating, John, 85
Kennedy, Robert, 242–3
Kerry, number of medical
    practitioners in, 65
Kickham, Charles, writer, 121
Kilkenny, 181

Killybegs dispensary district, 66
King, Archbishop William, 15, 17, 22,
    27, 28
Knights of Saint Columbanus, 208,
    211

Lady Dudley Nursing Scheme, 59, 69
    and handywomen, 71
    and travel difficulties, 69
League of Nations, 213, 214, 215,
    216, 217
Leitrim, county, 152
life expectancy in Ireland, 231
    compared to Europe, 231–2
    female advantage, 241–2
    improvement in, 239
Limerick district asylum, 105
Liverpool, 107
Local Government Act (1898), 52
Local Government Board, 141, 143,
    151, 192
'lock' wards, 194
Lockhart, Dr C., Sussex lunatic
    asylum, 103
Longford, board of guardians, 154
Londonderry county council, 194
Lord Lieutenant, 98, 107
Loudon, Irvine, 68, 69
lunacy, 95
lunatic asylum, admission to, 86–7
lunatics, 96
Lynch, John, coroner in county
    Clare, 83
Lyons, J.B., 115, 118, 119, 127–8
    writings on Sigerson, 127, 128
Lynn, Dr Kathleen, 151, 156, 209

MacCabe, Dr F.X., 93, 107
McAllister, Sarah (infanticidal
    mother), 105
McAvoy, Sandra, 186
McCarthy, Bishop, 81
McCarthy, Michael J.F., 105
McClelland, Richard, MRCSI, 63
McCullagh, James, 81
McDonnell, Ann (infanticidal
    mother), 104
McHugh, Bishop, 197
McHugh, Susan, 83

McNamara, Dr George, 86
McPolin, Dr James, 220
McQuaid, Archbishop Charles, 221
Madden, R.R., 114, 131
Maginn, Thomas, 81
Malcolm, Elizabeth, 2, 3, 4, 5
Malone, M.J., dispensary doctor,
    reminiscences of, 62, 69
Mandeville, John, death and
    inquest, 49–52
Mansion House Relief Committee,
    123, 130–1
maternal health, 211, 215, 217, 219,
    243
maternal instinct, 93
maternal mortality, 234
Maudsley, Dr Henry, 101
Maynooth college, 81
Mayo, suicides in, 83
Medical Charities' Act (1851), 4, 58,
    60, 67, 72
medicines, 29–30
medical history
    future of, 1–2, 4
medical practitioners, 22, 23
    in rural Ireland, 58, 49
    registered, 64–5
    unregistered, 70
Medical Registration Act (1858), 60
medical system, 14
medical witnesses, 42, 47
    disputes over, 42–4
menstruation, 101
Milltown, dispensary, Mullingar, 63
Milne-Edwards, Henry, zoologist,
    120
Ministry for home affairs, 194, 195
Mitchel, John, 118
Mitchelstown, county Cork, 49,
    143
Modern Ireland by George Sigerson,
    122
Monaghan Poor Law Union, 67
Moorehead, Dr, 155
Moorhead, Dr George, 51
mortality,
    and illegitimate children, 235
    factors affecting, 236–7
    in Ireland, 230–6

maternal, 234
    rural, 236
    urban, 236
mother and child scheme, 220,
    245
mothers, mental state of, 100
Mullin, Dr Peter, 64

Napoleon III, 118
National Children's Hospital, 209
National Council for Combating
    Venereal Disease, 196, 198
National nutrition survey, 237
Neilson, Samuel, 115
neuropathology, 119
Neville-Rolf, Mrs, 197, 198, 199
New Ireland Review, 120
Newcastle dispensary district, 63
Newman, John Henry, 123
New Ross poor law union, 66
Newport, Sir John, 95
North Dublin union, 88
Northern Ireland, 191
nurses, 59
nutritional standards, 237–8

O'Casey, Sean, 150
O'Connor, Charles, of Ballanegare,
    129
O'Daly, John (poet), 116
O'Flaherty, Mary (infanticidal
    mother), 104, 105
O'Halloran, Sylvester, 114, 130–1
O'Kelly, Sean T., 214–15, 216, 219
O'Mahony, Maud, 87
Offaly, flu in, 159
Offences against the person act
    (1861), 93, 94, 178
Origin of Species, 117

papal encyclical, 211, 213, 215
Paris, 118
Parsons, Sir Laurence, 16
patent medicine, 13
patients, 14, 23
    and dispensary doctors, 68
    and distance from dispensaries, 66,
        68, 69–70
    in Dundrum Asylum, 95, 96, 97, 98

Perceval, John, 24–5
physicians
  in Dublin, 18, 22, 24, 30, 38
plague, 15
Poor Law Commissioners, 58, 60,
  63, 68
Porter, Sir George, 43
Post mortem examinations, 39
Poynings law, 80
*Proceedings of the Royal Irish
  Academy*, 116
prisoners' petitions, 80
Privy Council, 15
prostitution, 100
psychiatrists, 108
psychiatry, 93
Public Health Bill (1947), 219, 220
Public health committee, 201
Public Health (Medical Treatment
  of Children) (Ireland) Act
  (1919), 220
puerperal insanity, 100, 180
puerperal mania, 102
puerperal sepsis, 234, 239
purgatives, 27
purges, 27

Rainey, Margaret (infanticidal
  mother), 98, 102
Ranelagh and Rathmines,
  Dublin, 64
Rathdown poor law union, 63, 66,
  153
religion and illness, 147
religious mania, 105
report of the committee on evil
  literature, 210
*Report as to the Practice of Medicine
  and Surgery by Unqualified
  Persons in the United Kingdom*
  (1910), 70
Revington, Dr, 98
Reynolds, James, 29
Ridley, Dr James, inquest, 51–2
Roche, Dr John, 146
Ronvier, Louis-Antoine, 119
Rosenberg, Charles, 160
Rossiter, Dr Thomas, 64
Rotunda Hospital, 212

Royal College of Physicians, 18
Royal College of Surgeons, 19, 20,
  30, 60
Royal Commission on Capital
  Punishment (1866), 102
Royal Commission on University
  Education, 117
Royal Commission on Venereal
  Disease (1916), 191, 192
Royal Medico-Psychological
  Association, 101
Royal Prisons Commission
  (1884), 123
rural Ireland, 57
  gender imbalance, 244–5
  medical practitioners in, 58
  mortality, 236
Russian flu, 141–3, 156, 158
  in Ireland, 143
  symptoms, 143–4
Ryan, Dr James, 219

Sadlier, Mrs (murderer), 105
Salpetriere, Paris, 119
sanitation, 240
sanitarian ideas, 155
Scotland
  and venereal disease, 195, 200
Scull union, 153
sex education, 221
Sigerson, George
  and intellectual context, 132
  and medical mentors, 120
  and republicanism, 122
  contributions to journals, 122
  journalism of, 122
  memory of, 126–7
  Parisian connections, 121
  patients of, 121
  pseudonym, 'An Ulsterman', 122
  schooldays, 115–16
Sigerson Shorter, Dora, 114
Sinn Fein, 123, 151
Skibbereen, county Cork, 153
Sligo prison, 185
smallpox, 15, 16, 22
Smith, Rebecca, 95
Society for Physiology and
  Psychology, 120

Society for the Social History of
    Medicine, 3
Solomons, Bethel, Master of the
    Rotunda, 212, 217–18
spa, 28
Spring-Rice, Thomas, 95
St Edmondsbury, Lucan, private
    asylum, 86
St Ultan's Hospital, 209
Stafford Johnson, Dr J., 213–14, 221
Stopes, Marie, 212
Stopford Green, Alice, 125
Stokes, William, 121
Strabane, county Tyrone, 115
suicide, 79, 105, 184–5
    and court verdicts, 84
    and insanity, 80
    and law, 79, 80, 81
    and medical explanations, 89
    and temporary insanity, 79, 81
    convictions for attempted, 87
    felo de se (self-murder), 80
    in Mayo, 83
    rates of in Ireland, 1864–1920, 82
Sullivan, Hannah (infanticidal
    mother), 98–9, 102
surgeons, 19, 22
Sussex lunatic asylum, 103
Swift, Jonathan, 17, 22
Synge, Bishop Edward, 16, 25, 26,
    27

temperance, 148
temporary insanity, 51, 97, 180, 181
The Atlantis, 117
The Nation, 116
The Shamrock, 114
Thomson, Dr Charles, 198, 200,
    201–2
Tipperary and great flu, 152
Tralee hospital, 181

Tuam, county Galway, 154
tuberculosis, 157, 233
Tullamore gaol, 51
Tyndall, John, 117
Tyrone county council, 194
Tyrone hospital committee, 193

Ulster Medical Journal, 200
unmarried mothers, 183, 187

Vaccination Acts (1858, 1863), 4
Varian, Ralph, 116
Varian, Hester, 114
venereal disease, 5, 19, 26
    and army, 193
    compulsory notification, 200
    cost of treatment, 195–6
    in Scotland, 195, 200
    legislation, 192
    notifiable disease, 200
    publicity, 196, 197
    treatment centres, 192, 194, 198

Wakley, Thomas, 38, 48
Walsh, Michael, P.P., New Ross, 72
Walshe, Joseph, 215
Warwick university, 6
water therapy, 28
Wellcome Trust, 2, 3, 5
Wexford, great flu in, 144, 148, 153
    workhouse in, 64
White, Dr Francis, 96
Wilde, Sir William, 114, 121, 130
women who committed infanticide,
    96, 99–100, 102, 105
workhouse system, 144
Wynn, Catherine (infanticidal
    mother), 105

Yeats, W.B., 122, 132
Young Ireland, 118